T0093656

# GROUND CONTROL

## AN ARGUMENT FOR THE END OF HUMAN SPACE EXPLORATION

### SAVANNAH MANDEL

CHICAGO
REVIEW
PRESS

Published by Chicago Review Press Incorporated
814 North Franklin Street
Chicago, Illinois 60610
ISBN 978-1-64160-992-0


Library of Congress Control Number: 2024932178


Interior design: Jonathan Hahn

Printed in the United States of America
5 4 3 2 1

*To all the social scientists who fight for a cause.*
*I hope you get funding.*

"*To make a prairie it takes a clover and one bee, —*
*One clover, and a bee,*
*And revery.*
*The revery alone will do*
*If bees are few.*"
—Emily Dickinson, "To Make a Prairie"


"*We went to the moon to have fun,*
*but the moon turned out to completely suck.*"
—M. T. Anderson, *Feed*

# CONTENTS

# INTRODUCTION

# THE PROTEST

*King's Cross Station, London, 2018* CE

**It's easy to fall in love** with London. Easy to slip into its rhythm and cadence. To drift between pub and kebab shop, pause with a coffee in hand to watch a violinist perform at Covent Garden, or swim with a tide of pedestrians down Oxford Street. It's easy to succumb blissfully to that world, even with a hand clenched tightly to your wallet for fear it might be pocketed into the bag of another. And it's easier still, I think, to forget such a city's powerful position on the world stage.

But in the summer of 2018, London's political omnipresence and perfect positioning in what feels like, to many, the center of the world (though decolonial scholars would disagree) was the last thing on my mind.

I was simply happy to be back.

I had only just returned from my anthropological fieldwork, in which I had been studying human interactions with outer space at Spaceport America in New Mexico. A little sunburned and emotionally exhausted, I was blissfully unbothered by Earthly issues and far more focused on turning my collection of scribbled field notes into a cohesive argument.

And I was *struggling* to make sense of those field notes. Struggling to figure out how to balance what I had *seen* with what my participants had *said*. Because for some reason, the two did not match up. In one margin, I had scribbled comments about the struggles of Truth or Consequences, New Mexico—a city that funds Spaceport America via taxpayer money—with poverty and drought, and in the other margin I had a list of celebrities who had bought tickets for Virgin Galactic spaceflights. Pages of my notes were dedicated to protests at the French Guiana Spaceport and the Thirty Meter Telescope, and just as many pages questioned why commercial space companies strove to build more spaceports in other postcolonial regions.[1]

My participants' political mission statements and optimistic aphorisms uplifted a vision different from the commercial reality of the spaceport. These were the individuals I studied, and sorting through their beliefs and values about human space exploration made me realize that their dreams left many individuals behind on Earth. These dreams were also built on the backs of those who would never touch the stars.

At the time, my research at Spaceport America would not lead me too deep into this philosophical wormhole, and compared to the political conflict I was about to walk into at King's Cross Station, the internal dilemma of sorting through notes was far from existential. But my own internal crisis was preventing me from turning the wealth of experiential, qualitative data I had acquired in New Mexico into concrete, pragmatic analysis.

It wasn't just my notes that caught me off guard. Another aspect of the project had me on edge. Something that no adviser or handbook on ethnographic methods could have prepared me for: it was incredibly difficult to research what many viewed as a utopia.

People *liked* outer space (well . . . except for that one girl I dated who said that the thought of outer space gave her cold sweats), so to talk bad about outer space during my fieldwork felt akin to talking bad about someone's mother. As I started to uncover things that made outer space exploration (specifically *human* space exploration) look bad, it felt like I had little room to critique the prospect without alienating myself from half the world. What's more, the utopians living in the utopia I researched (a.k.a. Spaceport America and other commercial space organizations) not only didn't like to talk bad about outer space but also didn't like to talk about the children being sacrificed to make their utopia function smoothly. (This is a reference, in case you didn't get it, to an excellent short story by a science fiction icon.)[2] In other words, I had seen the skeletons hidden in the closet, and now, regretfully, I wished that I had interviewed them instead.

This was what was going through my head as I rode through the London Underground. Not politics. Not the massive, thousand-strong protest I was about to unknowingly walk into. No, I was thinking about what Lady Gaga was going to wear to outer space and how to carefully frame my participants in a light that didn't make them look bad.[3] I was thinking about how I would pull my literary punches and what cultural critiques I would mask behind a facade of empiricism. And I was trying to ensure that my participants' world looked utopian, while presenting a "scholarly critique" easily defendable for the purpose of earning a degree.

Intensely distracted, I did not even lift my nose from my field notes as the train pulled into King's Cross Station. I walked straight into a revolution.

———————————

I've always been able to walk while reading, and yes, many individuals—strangers, family, friends, teachers—have reprimanded me for it over the years. But strangers I could (and did) ignore, and so I walked, head down, nose in notes, through the maze of underground corridors that I knew better than the hallways of my own apartment building.

Except this time as I did so, something odd happened. The flow of passengers coming on and off the various trains in King's Cross Station became congested and then tangled, as if someone was redirecting everyone. Which was a feat in itself, seeing as King's Cross hosted over 140,000 passengers a day on average. I remember I sighed, brushed my blue hair out of my face, placed my book in my bag, and thought two things: One, that a train or an escalator must have broken down or perhaps that someone had thrown themselves onto the tracks. And two, that this was annoying.

Now, my words might sound heartless, especially if you know the context of what was happening during the spring and summer of 2018 across London. But this story is one of hard truths and unfiltered realities, in the context of both my personal journey as a scholar and young woman, and the development of the commercial space race. I refuse to censor my own thoughts because I want you, the reader, to see how my personal imaginaries evolved with the rise of the commercial space industry itself.

But I'll get to that in a minute.

A transport official *was* directing the crowd, and I remember them telling us (as if nothing in the world was wrong) that we couldn't exit the way we usually would have.

If you've never been to King's Cross Station, then you might not know that it has many, many exits. Once you exit a train and leave the ticket turnstiles, you enter what could only be described as a shopping plaza that is also connected to St. Pancras Station. Parts of it are aboveground and parts of it are underground. It is a massive maze of corridors and coffee shops and station entrances and exits, covering over seventy-five thousand square feet.

The crowd of passengers and I were redirected a second time . . . and then a third . . . and then a loud noise ruptured the subtle din of confusion. Yelling and screaming. Another heartless thought went through my brain: *There must be a drunk person causing trouble. Or someone who's mentally ill.* But it did strike me as odd that all of the passengers near the trains at that point were redirected *again* away from the source of the noise.

And then *again.*

And then they herded us like confused cattle into the above-ground section of St. Pancras Station—which has glass windows and doors on all sides—and I looked outside and realized that there were police lined up around almost all of the entrances and exits in an effort to block off a sea of protesters from entering the area.

The protesters were calling for an end to the Turkish invasion of Syria and the mass murder of Kurdish people.

And here I was caught up daydreaming about sending humans to space and what Lady Gaga would wear on her Virgin Galactic flight.*

Suddenly, something about that felt very, very wrong.

I stopped walking and stood absolutely still, staring out at the mass of people.

And as I did, members of London's typically subdued police force wrestled a particularly passionate protester to the floor right next to Platform 9¾.

In all my time in London, I had never seen a police officer put their hands on someone, and trust me, I worked in a pub and was well used to late-night drunken shenanigans. On the eve of the World Cup finals, our bar wasn't even legally allowed to serve beer in real glasses because of the prevalence of bar fights. Plus, I come from America, the land of police brutality protests, yet still it was shocking to see someone get tackled via brute strength next to an iconic Harry Potter–themed tourist destination.

I continued to stare, causing the line of disgruntled passengers being shepherded out of the station to come to a shuddering halt.

Someone pushed me, bumping my shoulder as they shoved past.

I exhaled. How long had I been holding my breath?

It was as if roots had sprouted from my feet and grown straight down through my boots. As if Mother Earth herself had turned my head and said, "Look. *Look, you fool.* They will be left behind. What

---

* No shade to Lady Gaga—she is a fashion icon.

say have they had, in these conversations about space, child? You're dreaming of the stars while *this* is happening?"

We, public commuters, weren't forced to stay in St. Pancras, and let's be honest, that would have made bigger news than the protest itself. Instead, transport officials kept the rear entrance farthest from the protestors open. But rather than leave and escape the conflict, I walked up to the second floor of the station, sat down, and stared out the window at the individuals packed together, holding signs, marching the streets of London.

It started to rain.

My field notes hung limp in my hands.

Who *was* really benefiting from human space exploration? How many other protests were happening around the world at that moment? What problems were facing Earth then, at the very moment I was daydreaming about creating colonies on Mars and sending tourists into low Earth orbit? And what was I, an anthropologist, doing researching outer space? Weren't we supposed to dig up artifacts and study villagers in a rainforest? Or maybe solve murder mysteries with forensics? Why was I helping any space capitalists hide skeletons in their closets?

This story starts at a moment of rupture. A night when the curtain was ripped down and an alternative future behind the veil exposed. I'm speaking in metaphors, of course, as those who write about human space exploration often do, but let me pause before continuing and make myself very clear: there are no utopian visions or interstellar daydreams or biographies of "rock star" billionaires in this story about space. This is a story that is as raw as it is grounded. It doesn't leave out the awkward small talk or the inappropriate jokes made over happy hour cocktails. These words are messy. They are human.

In this story I trace the commercial space race (as many have before), but rather than focus on its rise, I describe what might arguably be its downfall. I make an argument for an alternative future. One that includes the voices seldom invited into techno-scientific conversations.

Here, I investigate the construction of global priorities and beliefs as "the global" becomes increasingly local. I do so through a realm of expertise as new as the commercial space race: outer space anthropology. It is through this lens of expertise that I confront current and historic perceptions of human space exploration. I will explain the birth of this field, and its impact on space science, in the coming chapters.

Who really benefits from human space exploration when access is granted only to those who have the money to get there in the first place? Should humans explore space at all? Is it worth it to send humans to space? What cultural outcomes will result from continued human space exploration and the colonization of other worlds? And last what can we learn about our present selves by studying our most extreme visions of the future?

This confrontation begins at King's Cross in 2018, but it will travel back and forth through time to examine moments in our history and the present that act as analogues for our future.

To understand what motivates our species' drive to explore—and our most imperialist urges—I'll examine colonial histories and inequitable social structures stretching from ancient Greek history to the present day. I'll resurrect often unacknowledged Apollo-era protests and dissent, which are reflective of space ethics and protests that circulate on social media today. From there, this story revisits historic moments such as the invention and failure of the Concorde and Aramis and the exploration of Antarctica, as a way of understanding technological ventures such as Virgin Hyperloop and Virgin Galactic.

*Ground Control* will also shine a spotlight on congressional discourse, annual budgets, fictional representations, the narratives promoted by leading commercial space companies, space law, and social media, to tease apart answers to the questions posed above.

Sometimes this story is full of love and hope. Sometimes it's full of anger and resentment. This is how I was trained to write as an anthropologist: not to avoid bias but to be up front with it. To acknowledge emotion and reflexivity as part of the story. Which is why I start with a moment of personal transformation.

Was this the moment I transitioned from a firm believer in and advocate for the space industry to someone who couldn't shake a guilty feeling that shadowed the prospect of human space exploration? Not exactly, as you'll see. Like most personal transformations, this coming-of-age is akin to a botanical creeping. Vine-like tendrils pried away celestial concrete one brick at a time. An Earthly growth, sometimes painful, sometimes breathtaking.

But make no mistake—I want to go to outer space.

I want to see humankind construct cities on far-off planets and moons.

That doesn't make it right, though, and it doesn't mean that *now is the time to do so.*

What humankind is faced with are two very different utopian demands.[4] The first is the demand to explore outer space. To colonize. To ensure humanity's future. To search. To learn. To acquire new resources and conduct research. The second is the demand *to stay, to conserve, to repair, to care, to maintain, and to sustain.* This alternative vision of the future flows from the mouths of caretakers and conservationists—individuals who seek out restoration of the planet our species was designed for. The Caretaker's Demand argues that the money spent on advancing human space exploration might be better spent focusing attention back on planet Earth. On realizing post-scarcity futures, socialized health care, a world without territorial claims-making or colonialism, and other Earthward-focused demands that I'll describe throughout this book.

I can't deny that coming to terms with the prospect of not exploring outer space is hard. The startling idea that human space exploration might not be the best decision for mankind destroys the part of me that yearns for voyages into the unknown. A restlessness creeps in, not unlike a restlessness I felt through the COVID-19 quarantine. As someone who has researched the space industry for several years now, who has engaged deeply with speculative ideals, the prophetic, and the imaginary, and as someone who intensely loves the legacy of science fiction, I struggle with letting go of some of these more transcendental dreams. I also recognize the social fallout that can occur when making such a decision. If I side with utopian demand number two—the Caretaker's Demand that asks us not to explore the unknown but to focus attention on socioeconomic problems here on planet Earth—there go my prospects of working in the space industry; there goes my LinkedIn network. If my anthropological fieldwork taught me anything, it was that arguing against human space exploration would alienate me from the space industry.

But we have to make tough decisions as adults, don't we? We have to choose to tell the stories that go untold, to share the memories that might get us blacklisted, to stand up for the perspectives shadowed by the legacies of billionaires, to address racism, sexism, ableism, and colonialism. And the decision to leave Earth is one our entire planet is facing as we get closer and closer to the goals shared by so many commercial space companies.

But it is a decision that very few get a say in.

What would it do to humankind's collective psyche if we let the dreams of expanding human civilization into space go? In November 1998, the first module of the International Space Station (ISS) was launched into space. A decade prior to that, Space Station Mir was already maintaining a human population in space—beginning in March 1986 until it was decommissioned in June 2000. Which means that for the entirety of my life, human beings have been up there, looking down on us.

My mother used to wake us up to go trace the ISS with our fingers as it passed above our home, and I used to think, *There they are . . . astronauts. Always watching. Always there. Always above us.* Now there's talk of the ISS's imminent demise—due to its age, the maintenance it requires, and its expensive upkeep—and its replacement with a commercial space station. Though the prospect of the

ISS's end shakes me, I'm left wondering what would happen *if we didn't replace it*. What if, like any other bad, crumbling, expensive relationship . . . we let it go. And instead of jumping into another bad, crumbling, expensive relationship, what if Mother Earth and her beautiful inhabitants just stayed single for a while and focused on themselves?

What if we stopped seeking out poisonous atmospheres and bad relationships with foreign bodies, and focused on the wreckage in front of us? That overpopulated, polluted, absolutely decimated planet we were born into. What if we focused not on the space-faring generation but on *every* generation? What if Earth became our trillion-dollar project? What if instead of looking up and always begging for more more more, we stopped, accepted what we were given, and spent all that pent-up passion and wonder and curiosity on harnessing the deepest, hottest hydrothermal vents, the blistering desert sun, windswept northern oceans, and frozen tundra? What if we found new ways of existing in the harsh places we've ruled out on our home world before forcing ourselves into the ecosystems of others? The depths of the oceans, the vast plainlands, and the Great White North. What if we revisited the ways of the ancient seafaring cultures and the desert nomads? Drawing inspiration not only from the speculative but also from the real, the here, the now, the pragmatic and the historic.

What if we focused on *unmanned* space travel and let the last astronaut go?

Just like Assata Shakur's call for collective action—which seeks not only to reform but also to revolutionize and works to upend current systems that fail to value Black lives—the Caretaker's Demand must conjure up a vastly different world without apologizing for radical social changes.[5] A world where planet Earth takes priority in research, and the collective mission to ensure the sustainability of the human species works from home.

Consider this story a call to arms.

# PART I

# A LETTER TO THOSE LIFTING OFF

# 1

# PAST THE POTATO AND INTO THE FUTURE

*The Basement of Turlington Hall, Gainesville, FL, 2016* CE

The corridors of Turlington Hall's basement are lined with bones. Glass cases, filled with artifacts from eras long past, frame the corridors and office walls. Posters and magazine covers are proudly displayed on doors of faculty spaces and classrooms, paying homage to professors and students with recent publications or upcoming events. I think the decor was meant to make the department look cozier, and to show devotion to the field, but I'll admit I always thought the anthropology department at the University of Florida looked a little grim.

"It reminds me of an asylum," I told a friend once, during my first weeks on campus. And it did. The windowless basement corridors were divided up by large, swinging double fire doors, and several laboratories were dispersed throughout the hallways, home to hunched-over students in stark white lab coats. Once, before class, someone casually rolled in a cart laden with casts of the skulls of extinct hominids. We, anthropology students, would measure them down to the millimeter, rolling them gently in our hands beneath mesmerized eyes. One day I would be able to tell the difference between *Homo neanderthalensis, Homo habilis,* and *Homo sapiens* by a single tooth. A party trick that's unfortunately very difficult to show off.

My friend shrugged. "Anthropology departments always get the short end of the stick. Funding and all that."

Their words ring true, and I'm sure reverberate with annoying clarity in the minds of every social scientist or historian reading this book, but funding was the last thing on my mind on that breezy November day. At that moment, I was far more concerned with the fact that I was very late for a meeting to discuss my personal future in the field of anthropology.

I didn't realize I was late until I passed the twenty-five-million-year-old rock from the Oligocene Epoch (known fondly by the Florida Gators as "the potato") that sat outside Turlington Hall. As soon as I checked my watch and saw the minute hand drifting past the five-minute mark, I picked up my pace, swinging my backpack over my shoulder and taking the stairs two at a time.

"Shit. She's gonna kill me," I mumbled to myself.

As I scurried down the hall, Dr. Peter Collings passed me, shuffling along in seal fur moccasins given to him during his fieldwork with the Inuvialuit peoples of the Canadian Arctic. Seconds later I barely avoided running into a grad student carrying a precariously stacked pile of books in their arms as I tried to wave hello and maintain my pace. Collings just shook his head, chuckled, and waved back as I disappeared around a corner.

Little did I know that I was taking the first steps toward following a legacy of anthropologists and other social scientists who were researching outer space. Neither I, nor the professor I was charging down the hall to talk to, knew this small group of cultural experts on outer space existed, but just years prior they had begun their own adventures, initiating what would be the birth of a subdiscipline. For example, Valerie Olson, environmental anthropologist and expert on extreme environments, had worked on the NASA Extreme Environment Mission Operations (NEEMO) project on the bottom of the ocean floor.[1] Stefan Helmreich had studied the probabilities of extraterrestrial life by examining undersea hydrothermal vents that shouldn't have supported life at all.[2] Janet Vertesi, science and technology scholar, had just spent several years working with scientists who were building the *Spirit* and *Opportunity* rovers to understand how humans anthropomorphize and instill human values in inanimate space objects.[3] Sarah Parcak, TED Prize-winning archaeologist, collaborated with NASA to use satellites to locate uncovered pyramids in Egypt. And Alice Gorman, known fondly as Dr. Space Junk, used her skills as an archaeologist to study satellite debris.

My story was about to intersect with theirs.

But in those days, it felt impossible to research outer space as an anthropologist. And I sensed my personal interests in speculative futures, science fiction, and space exploration shearing away from the fault lines of my chosen field of study. How does an anthropologist research outer space? Had I applied for the wrong major?

Don't get me wrong, I adored anthropological research of all types. Linguistics. Archaeology. Evolutionary studies. Social anthropology. Forensics. Egyptology. Classics.

By the time I was in third grade, I had it in my head that I would be an anthropologist and declared it to anyone who would listen. Part of the reason for my odd fascination with the field at such an early age was because I was exposed to it so heavily as a child. My parents—an accountant and a programmer—just happened to have a passion for history and culture, which they shared with my sister and me throughout our childhood. We visited archaeology sites, museums, war memorials, Amazonian villages, Scandinavian cultural sites, Central American pyramids, American Indian burial mounds, and other historic sites across the globe. On top of that, I was obsessed with ancient languages and tried to teach myself ancient Mayan, Egyptian hieroglyphs, ancient Chinese, and fantasy languages such as Dinotopian and Elvish (I remember only one of these languages today, and it's perhaps the least useful of the bunch: Dinotopian).[4]

Anything my mother read, I read, which at the time was a lot of classic literature, history books, and scientific magazines.

I was privileged enough to have parents who were genuinely supportive of my obscure interests. If I became interested in Norse deities, they took me to the library and helped me check out books on Norse deities. When world war history fascinated me, my dad and I sat down and watched *Schindler's List*, *Saving Private Ryan*, and a long list of related documentaries. But between books about atomic bombs, Celtic fables, and ancient Viking boat building, I was also reading what every other kid was reading at my age. Adventure novels, science-fiction stories, and cheesy romances.

By the time I was in high school, my interest in science fiction had peaked. Gone were mummies and ancient alchemy. Now I was daydreaming about becoming an astronaut and designing rocket ships. But the prospect of a career in space science felt impossible. I was terrible at mathematics and never passed calculus or physics. There was no way I could have a career in space! Let alone combine my obscure interests in cultural history with such a techno-scientific field.

Right?

*Right?!*

Yet there I was, charging toward Dr. Susan Gillespie's office, about to ask her if it was possible.

I arrived, breathing heavily, with my hands on my knees in an attempt to compose myself before knocking on her door. Unfortunately, she must have heard my anxious scrambling, because the door swung open before my hand even touched it.

"Good afternoon, Miss Mandel," she said, looking down at me over wide-framed glasses.

I gave her a guilty grin and smoothed down my hoodie. The action did absolutely nothing to rid it of what were permanently ingrained wrinkles.

Something about Dr. Gillespie reminded me of Professor McGonagall in the Harry Potter series. She was a strict teacher, but I spent the entirety of her classes on the edge of my seat in contented fascination, and throughout the process of writing my undergraduate honors thesis on the cultures of restaurants, we became close.[5] I specifically admired the way she looked at the world, which is why I sought her advice on choosing a graduate school.

"Well, what are you interested in researching?" came her first question as she settled behind her desk. Strewn across its wooden frame was memorabilia from her fieldwork on pre-Columbian Meso-american cultures.

"I'm really not sure."

"Social anthropology though, right?" This question was paramount, for reasons I'll explain now, before going forward.

Anthropology is divided into four primary subfields: Social or cultural anthropology, archaeology, linguistics, and biological anthropology. Those subfields are divided even further. For example, biological anthropology includes: primatology, paleoanthropology, paleopathology, evolutionary anthropology, genetic anthropology, forensics, and others. Social anthropology includes: gender studies, material culture studies, environmental anthropology, political anthropology, and more. And a lot of these subfields tend to overlap. For example, an archaeologist could do genetic studies, or a linguist could specialize in evolutionary anthropology by studying the evolution of the tongue or primates and language. When it comes to anthropologists who research space (which I'm getting to), there are "space anthropologists" from all four of these subfields.*

---

* Space anthropologist was a term given to my profession by Kris Kimel, cofounder of Space Tango. It started gaining popularity after *Ozy* magazine released a profile using the term to describe my research area. As far as I know this was the beginning of its usage, though the term *space archaeologist* has been around longer, in reference to research conducted by Alice Gorman, Sarah Parcak, and others. (Leslie Gutman, "A Space Anthropologist Warns: Inequality Gets Worse on Mars," Ozy, October 24, 2019, https://www.ozy.com/the-new-and-the-next/a-space-anthropologist-warns-inequality-only-gets-worse-on-mars/221963/ [page discontinued].)

Parcak and Gorman are both archaeologists. They adopt the tools of their subdiscipline and use them to research historic perceptions of outer space by studying material culture. Similarly, there are evolutionary anthropologists who research extremophiles and how the human body might evolve in space, social anthropologists (like me) who study what it means to be human when we keep pushing the boundary of humanity, and biological anthropologists who study the future of sex and reproduction in space.

I wanted to be a social anthropologist because, in my mind, social anthropologists could research *anything from anywhere*. Board game geeks in Belgium, mobsters in Milan, marketplaces in Mumbai, fashion in France, raves in Romania, you name it. The doors were wide open. At least, that's what I thought.

"Yes, social anthropology," I responded tentatively.

Dr. Gillespie nodded, her gaze thoughtful, "Well, what are your interests? What do you like?"

"Uh . . . science fiction? Outer space?"

Her typically calm and collected expression wrinkled as she began to laugh in an easygoing way. "Tickets onto the ISS are a bit out of a graduate student's research budget, don't you think?"

And that was the problem. Social anthropologists, *almost always*, have to do fieldwork. Typically ethnographic fieldwork, which involves *living* with the culture you're studying, or at least interacting intensely with them. How on Earth (pun intended) was I supposed to "interact intensely" with outer space? What was I going to do, live on the ISS as a researcher? Hang out with astronauts all day? Anthropologists have trouble getting a small summer research grant, let alone a multimillion-dollar ticket onto the ISS. And when I say *live*, I mean spend *years* with a culture, repeatedly returning to them summer after summer (because most anthropologists work on academic schedules). Sure, spending a couple months with a culture is also acceptable, but when it comes to traditional anthropology, spending *years* is far more typical, and relationships with field sites linger for entire careers.

For instance, Collings, who I passed in the hallway at the beginning of this chapter, has conducted research with the community of Ulukhaktok in the Canadian Arctic since 1992. And there's a very good reason for this sort of extended time frame. To gather meaningful qualitative cultural data, one typically has to bond with a culture's indigenous population (in the case of my research this would be those who interact with outer space) to the extent that they are willing to share social intricacies and experiences about their culture that one can't learn simply through observation or literature. The researcher must be *invited in*—which takes many forms and can mean many things—and this takes time. Lots of time. Sometimes it requires learning a language or multiple languages. For example, famous anthropologist Clifford Geertz wasn't accepted by his participants until he ran from the police with them during a raid on a cockfight.[6]

Dr. Gillespie must have seen how crushed I was by her response, because her gaze grew sympathetic. "I'm just not sure it's possible," she said gently.

Not only did it seem impossible to choose "space" as my field site, but astronauts very rarely plastered their contact information across the internet. Where was I even supposed to begin when it came to getting in contact with someone who had been to outer space? These difficulties went beyond logistics. Outer space exploration is incredibly speculative in nature. Sure, man had been to the moon and rovers to Mars and probes to Jupiter, but so much of what I was interested in hadn't happened yet. Colonization of other planets, resource mining of asteroids, space tourism . . . it was all just science fiction at this point. (My research began in 2016, at a time when Virgin Galactic was still recovering from the crash of the VSS *Enterprise*.) Space travel was not real, or here, or now.

*How does somebody research something that hasn't even happened yet?*

To give Dr. Gillespie credit, science and technology studies were completely out of her wheelhouse. Why should an academic trained primarily on Aztec, Mayan, and Olmec studies know anything about the anthropology of outer space? In general, academics tend to know a lot about one niche. And let me tell you, even today, forty years after outer space anthropology started to kick off with the publication of the edited volume *Cultures Beyond the Earth* and Peter Redfield's groundbreaking *Space in the Tropics*,[7] I still get emails weekly from tenured professors saying, "Woah. I had no idea anybody was researching this stuff. This is amazing."

The research on space that was being done at the time, that neither Dr. Gillespie nor I knew existed, was . . . obscure. Let's be clear, it still is obscure. But there *were* space anthropologists out there at the time of our conversation, and they were researching things that did exist, as well as things that didn't exist by localizing that research here on Earth.

For example, when Vertesi studied the *Spirit* and *Opportunity* rovers during their development, she found that human culture is extended into the stars via robotic entities through embodied practices and anthropomorphization. In other words, we leave behind a little bit of ourselves in probes, rovers, and satellites when we send them out into space. Sometimes without even realizing it. These inanimate technological objects aren't just hunks of metal created to serve a purpose. They become imbued with beliefs, values, politics, and emotions, and extend these social and political systems when we are not physically there to do so. Why did programmers have *Curiosity* sing "Happy Birthday" to itself? Why did an astronaut create a blog called "Diary of a Space Zucchini," which ventriloquized space vegetables for the masses?[8] Programmers, astronauts, scientists, and institutions use inanimate objects to extend political and social ideologies into outer space.

Now, the emotional part of this might be easier to understand than the political aspect. After all, when China's *Jade Rabbit* rover and the *Opportunity* rover went out of service, their "deaths" made tearjerking headlines for weeks.[9] But Lisa Messeri, another anthropologist whose work has taken a space focus, has studied the political aspect of the values we embed in inanimate objects related to outer space. Take, for instance, something as seemingly benign as Google Maps. Messeri argues that through the creation of extraterrestrial Google Maps (in other words, exploring Mars on Google Maps) scientists socially construct Mars as inherently "democratic" and "dynamic" as a way of expressing certain political goals.[10] Likewise, Redfield's work, which I mentioned earlier and will talk about with more depth in later chapters, is equally political as it interrogates the impact of a spaceport on a (supposedly) postcolonial colony. Redfield describes the constant political negotiation involved between scientific outposts and local populations.

In other words, human space exploration and techno-scientific development is never benign.

Beyond the inanimate, space anthropologists also study the future of human evolution and the potential of extraterrestrials, as well as the social and cultural dynamics of astronauts. Helmreich has worked at diverse locations, including Monterey Bay, Hawai'i, the Woods Hole Oceanographic Institution, and undersea volcanoes in the Pacific to study life that thrives in some of the most extreme conditions known to planet Earth. By examining life at the extremes, we can begin to understand how the boundary of the "extreme" is getting pushed further and further out into the stars, especially in the context of human space exploration. As these boundaries get shifted, so too do definitions of what it means to be human, along with conceptions of temporality and spatiality. Olson has also contributed intensely to this topic through her work with the NEEMO undersea laboratory.[11]

Likewise, Dana Burton and Debbora Battaglia have explored representations of extraterrestrials in literature and media, which need to be taken seriously in light of the conversations about science fiction.[12] Science fiction can teach us a lot about science fact, and indeed these forms of speculative media influence politics, law, culture, science, and technology. I'll give examples of this throughout *Ground Control*.

I left Dr. Gillespie's office feeling conflicted. Though anthropologists were engaging with outer space—across the globe—I was not yet aware of it. And I would not be until I left behind those early years of my academic life at the University of Florida and moved to a different campus, in a much bigger city: London.

Over the coming summer, while I tried to internally balance my love of anthropology and my interest in science, I worked on archaeology sites in Tennessee and North Carolina. The digging, I found, was both therapeutic and thought provoking. The long sunny days, the soil and muck, the gleeful howls of young archaeologists chasing a groundhog out of a sixteenth-century Native American site, the cleanse of the river nearby on lunch break, the digging and digging and digging—it certainly reenforced that I would have to maintain some part of this path as I moved forward. And it reenforced who I was at heart. I was an anthropologist.

# 2

# THE CROSSROADS

*University College London, 2017* CE

**When I first arrived in London,** I wasn't in love with it. In fact, I was miserable. Not even twenty-two years old and recovering from a divorce from my high school sweetheart (who turned out to be not so sweet), I found myself adrift and purposeless in a city too easy to feel lost in. And in parallel with such spiritless attachment to my Earthly surroundings, I found my star-studded academic dreams clouded by all the light pollution.

I *believed* Dr. Gillespie when she said she didn't think it was possible to research outer space as an anthropologist and so spent those early months earning my first master's degree at University College London (UCL) hunting for alternative research plans.

I considered researching the *Orient Express*, American stoner culture, cam girls, British pubs, and continued to feel more and more disconnected from my choice of career . . . until one day, by pure happenstance, I declared to one of my professors, Dr. Aaron Parkhurst, "I just wish I could research space!"

We were standing together chatting after his class on the study of sexuality and reproduction, and I'll never forget the absolute look of surprise on his face when I said those words.

"But *I* research outer space," he laughed, as if his response meant nothing profound at all. Then he began describing his research on the human body and long-term space exploration.[1]

Aaron studies the way the idea of the human body will be reconfigured as it becomes less and less equipped for the environment it's living in. In other words, how might "being human" and human functions change on Mars, and specifically in a closed-loop environment?[2] How might the concept of what it means to be an Earthling change? How might the uniquely constructed habitats of future Martians affect their bodies and bodily functions?

After patiently listening to me share my eager interest in researching space as a social scientist, Aaron said, "You really need to speak to Victor and Jeeva." He was referring to two other professors in the department of anthropology at University College London.

"They're developing a project that's going to work closely with the ISS."

I recoiled from him slightly. Aaron got so excited when he talked, and the two of us had this habit of gravitating inward as we spoke, like confidants spilling girlish secrets or binary stars caught in orbit around each other. It wasn't our proximity that took me aback though. It was what he said.

Dr. Gillespie's augury echoed in my mind: *"Tickets onto the ISS are a bit out of a graduate student's research budget, don't you think?"*

Shaking my head in disbelief, I opened my mouth, a thousand questions on the tip of my tongue. What were they doing? How were they researching outer space? Were they the only ones? What were their specialty areas? Were Victor and Jeeva biological anthropologists like Aaron? Social anthropologists? Had they actually spoken to astronauts or visited NASA? I needed to see their publications . . . attend their talks . . . ask them to be on my committee . . .

I shut my mouth, opened it again, shut it again, gaped like a fish out of water, and then took a deep breath. "Where are their offices?"

———————

Dr. Aaron Parkhurst, Dr. Victor Buchli, and Dr. David Jeevendrampillai's project, Ethno-ISS, received funding from the European Research Council under the European Union's Horizon 2020 program just after I finished my master's degree at UCL.[3] The project was awarded over €2 million to study what was arguably the only extraterrestrial society in existence.[4] Aaron's position on the project was to study astronaut interactions, embodiment, and emotion on the ISS. Other participants would research the effects of isolation, design, architecture, habitat development, and speculative policy.[5] Anthropologists were well and truly studying humanity in one of its most inhuman forms. The Earthling, *sans Earth.*

But my own cosmic fieldwork took me not to space (at least not directly) but to the desert.

Like I said previously, the Ethno-ISS project was funded *just after* I left UCL, leaving me to come up with a field site of my own, at a time when no previous social science project on space had been awarded millions of dollars in research funding (though space archaeology work, such as that done by Parcak and Gorman, had received funding from space organizations previously). In other words, getting potential field sites to take my project proposal seriously was a *nightmare* fit for the pages of a just-as-discriminatory Lovecraftian-style novel.

But I'm getting ahead of myself again.

How does one *choose* a field site?

I've already discussed the importance of spending extended time with a field site and being accepted into the culture one is studying (something that other anthropologists have commented on in far greater depth),[6] but how does an anthropologist gain access to a field site in the first place? Especially when the field site of interest is a highly restrictive location, often *literally* hidden from the public, and covered in more red tape then Kim Kardashian's Paris Fashion Week Balenciaga catsuit.

"Gaining access to" and "choosing" a field site are two very different things and involve two very different processes, albeit interconnected ones. I'm of the strong opinion that for every iota of agency a researcher gets in the selection process, their participants get a thousand times that. No matter how much funding one receives, nor how accessible a field site is, participants must *agree* to be studied. Power is (and should be!) in the hands of those being researched. Whether they are indigenous members of a remote tribe or space capitalists. This is why institutional review boards exist and why almost all ethnographic projects have to go through IRB approval. My master's dissertation certainly did.

Thus, gaining access to and choosing a field site are complex and sensitive procedures, accomplished in diverse ways. For example, environmental anthropologist Valerie Olson gained access to her field site—NASA's Johnson Space Center and eventually *Aquarius*, NASA's underwater research station—by becoming a research intern at the National Space Biomedical Research Institute in 2005. It took her over three months to officially get badged in.[7] Over time, her position as an intern led to acceptance as a researcher. Through the support and acceptance of her participants, Olson ended up conducting research at the NASA Ames Research Center, NASA Headquarters, the European Space Agency, and several conferences—such as those hosted by the American Astronautical Society and International Astronautics and Aeronautics Association.[8]

Social anthropologist David Valentine also saw value in conferences as field sites, to the extent that they became the main site of his research on commercial space industry perspectives of profit, temporality, and risk.[9] After several failed attempts at developing working relationships with SpaceX and smaller commercial space companies, Valentine sought out commercial space industry conferences believing they would be an excellent location to network with potential field sites. And they were! Once in attendance, Valentine realized that he did not need to look much further than the conferences themselves. They were a central meeting point for dozens of companies, and the tone of such events was one that welcomed the inquisitive investigations of an anthropologist interested in outer space. Eventually his field work would expand to include networking events at the Mojave Air and Space Port, as well as online discussion forums.

---

In my case, finding a field site involved sending emails and making phone calls to commercial space companies, government-contracted space projects, astronauts, science-fiction authors, and museums.

A little over *three hundred* emails and phone calls . . .

. . . which I kept track of in one giant Excel spreadsheet.

As it turns out, many astronauts do simply have their contact information posted online.

In total, I received maybe a half a dozen replies—most of which were from people genuinely interested in what an anthropologist was doing researching outer space but who either couldn't allow me to conduct research with them due to existing funding schemes or confidentiality issues or didn't see the point in it because my research methods were largely qualitative. Looking back on it, I sympathize with perspectives like these. It wasn't until late into my graduate program that even I truly understood the potential of qualitative anthropological research and its ability to have massive political, economic, medical, legal, and social ramifications.

But the result of qualitative anthropological research can be very explicit. Take, for example, a study done by Teun Zuiderent-Jerak on hemophiliacs at a hemophilia care center.[10] Zuiderent-Jerak conducted ethnographic research on the discrepancy between the treatment that the hemophilia care center was advising and the actual treatment patients were performing at home. In other words, patients weren't taking their medicine correctly or on time, and they were having dangerous bleeds or, in worst case scenarios, dying. After spending extended time with those individuals and coming to learn about their personal practices, beliefs, and values related to their medication, Zuiderent-Jerak started working with the clinic hematologist to devise a technological intervention to help patients maintain advised and timely use of their medication. Zuiderent-Jerak also helped the clinic establish a multidisciplinary hemophilia consulting hour when patients would speak with a nurse directly to discuss possible events in their daily lives that might alter the use of their medication.*

My correspondence with potential field sites was often spent telling stories, such as Zuiderent-Jerak's, to try to convince them of the value of qualitative anthropological research and the role it could (and would) play for space exploration.

———

Several months into frantic callbacks and voicemails and unread messages, Jane Kinney at the Commercial Spaceflight Federation connected me with Karen Barker, a now-former director of Spaceport America.

I was sitting in my dorm room in London, on the edge of my less-than-twin bed, when Karen called. My room was on the tenth floor of an apartment block on Caledonian Road in North London. The view was extraordinary, and wild poppies grew on the rooftop below us every spring, but the elevator had a habit of breaking, often with me inside, so I was happy to be rid of it when I finally moved into an apartment on Regent's Canal.

Karen was equal parts hesitant and eager to have me conduct research at Spaceport America. She was very curious about my work

---

* Further research by Zuiderent-Jerak specifically debates the role of an ethnographer when it comes to intervention. See Teun Zuiderent-Jerak, "Blurring the Center: On the Politics of Ethnography," *Scandinavian Journal of Information Systems* 14, no. 2 (2002): article 9.

and interested in having me come out to the spaceport. But I sensed a reluctance to her invitation and knew it had nothing to do with the usual skepticism about my field of study. Karen was worried about current events.

I connected with Spaceport America during a very transitional, sensitive, and restorative time in their development. They were still recovering politically from the crash of the VSS *Enterprise* in 2014, which resulted in the death of test pilot Michael Alsbury. Alsbury had been involved with Scaled Composites, the company building Virgin Galactic's spaceships since 2001, and he and his copilot, Mark Stucky, were treated with a reverence akin to Apollo-era astronauts. One recent publication by Nicholas Schmidle went so far as to compare the two to deities, terming them "test gods."[11] So during Spaceport America's Artemis-era* ventures, headlines like *Wired*'s "Space Tourism Isn't Worth Dying For" still haunted Spaceport America.[12] In addition, the spaceport was suffering from a far more recent string of negative publicity that stemmed not from any specific incident but from what outwardly seemed like inaction in the eyes of the public.[13] What had the spaceport been up to since 2014? Where was the people of New Mexico's money going?

Actually, the spaceport had been up to quite a bit! But they were collaborating with commercial space companies focused on far less flashy launches, such as rocket launches, or working with other projects that required more confidentiality. As for where taxpayer money was going . . . well . . . there has been some question over that in recent years.[14]

But reporters were not the only ones to critique what seemed to be a plateau in Virgin Galactic's glamorous space tourism plans. During my ethnography, I found that the citizens of Truth or Consequences (the town closest to Spaceport America) were growing more

* "Artemis era" references the Artemis program initiated by the Trump administration. The term mimics "Apollo era" which refers to the period between 1963 and 1972 when the Apollo program was active.

and more anxious to see spaceships fly over their tiny town again, and I even had an astronaut rant to me in the months following my research that he really "wished Virgin would get their act together."

There was a sense of fragility that hovered over that town nestled in the Jornada del Muerto desert basin. A feeling that Spaceport America was one of the only things keeping it from "going ghost" and that it wasn't living up to its promises. One speculative documentary by Hannah Jayanti does an excellent job, I think, of showcasing the intense duality of hope and hopelessness that overshadowed the spaceport in this transitional period.[15]

This was the Spaceport America I was being invited to study. Not the 2021 Spaceport America, successfully hosting crewed Virgin Galactic flights. In 2017, Spaceport America was at a turning point as much as I was. Making it even more fitting that the city where their headquarters was based—Las Cruces—directly translated to "the crosses."

I responded to Karen's hesitancy by explaining to her that anthropological research, no matter where it occurs, maintains *neutrality through positionality*. In other words, researcher bias should not be ignored, but instead the beliefs, values, and judgments of the researcher should be acknowledged and then set aside as one observes, learns, and interacts with the culture they are studying. This is because it's impossible to well and truly ignore one's own positionality! It can't and shouldn't be done.*

Ironically, my beliefs, values, and judgements about outer space and those who engaged with it coming into my research on the topic were comparable to idolatry and obsession. At the time, I would not have objected to worshiping Alsbury and Stucky as deities. Quite the contrary—I would have built them an altar.

---

* And before you say, "Well, this isn't a very neutral book, Savannah," keep in mind that this book is not a dissertation nor an ethnography.

As a child raised on the visionary, otherworldly words of Isaac Asimov, Kim Stanley Robinson, and Kurt Vonnegut, I wanted nothing more than to be a part of the world of spacefaring humans. In the years leading up to my fieldwork, I would have never thought to critique human space exploration because I didn't think there was anything to critique. I was a rabid believer in the utopian visions conjured up by space organizations. Human space exploration, in its most speculative and down-to-earth forms, evoked for me intensely emotional imaginaries of hope, adulation, inspiration, and exaltation. But like I said before, this was just the beginning, and we were at a crossroads.

# 3

# ON FAITH AND SACRIFICE

*Spaceport America, Truth or Consequences, NM, 2018* CE

**Initial conception of Spaceport America** began in the 1990s, overlapping with some of the first commercial rockets to be launched out of the neighboring White Sands Missile Range (WSMR). The Spaceport's proximity to WSMR, and its conveniently high elevation of 4,595 feet, was purposeful. WSMR offered Spaceport America something no other spaceport had: six thousand square miles of reserved airspace. There was no need to fight with commercial airlines over flight schedules. All commercial space companies had to do was coordinate their launches with WSMR and the Federal Aviation Administration and hope for good weather, and they were more or less ready to go.

By 2005 Virgin Galactic announced its plans to headquarter out of what at the time was being called the Southwest Regional Spaceport, and in 2006, the spaceport had been renamed Spaceport America. Scaled Composites, the company that had designed *SpaceShipOne*, Virgin Galactic's primary spacecraft, had recently won the Ansari X Prize in 2004. The $10 million prize was awarded to the first non-government organization to launch a reusable spacecraft into space twice in two weeks. Shortly after, New Mexico approved $110 million in state-supported funding to construct the Spaceport America facility that exists today. In these early years, Virgin Galactic predicted they would fly fifty thousand passengers during the first decade of operations using five SpaceShipTwo and two WhiteKnightTwo vehicles.[1] It was also proposed that ticket prices would start at $200,000 and then drop within a decade to just $25,000.[2] While in actuality Virgin Galactic has sold tickets to customers at three different prices: $200,000, $250,000, and $450,000.[3]

Over the next few years, residents of Doña Ana County, Sierra County, and Otera County would vote to assist in the construction and continued operation of Spaceport America through taxpayer money. From 2010 to 2021, Doña Ana County taxpayers paid $90.25 million toward construction costs for Spaceport America.

New Mexico legislators are now fighting to claim tax revenue from commercial space passenger tickets, which Virgin Galactic has loopholed their way out of by classifying passengers as "freight."[4]

The financial sacrifice that has been made by these surrounding counties, and the faith that the residents had in Spaceport America, was something participants brought up often throughout my fieldwork. Progress could not be achieved without penance, and in the eyes of my participants, Spaceport America was far more than an outlet for tourism or a gateway to low Earth orbit: it was a stepping stone to ensure human colonization of other planets, and eventually the future of the human race as a species.

---

I arrived in Las Cruces, New Mexico, exhausted from a thirty-hour, five-day drive from South Florida, already suffering from the effects of the altitude. I would spend the last part of my road trip traveling along Mexican border lines that I was warned never to cross over as a young, white female. And don't misunderstand me. I wasn't naive enough to truly consider a trip to Ciudad Juárez. The city and its surrounding areas have had long-standing travel advisories due to high rates of murder, kidnapping, robbery, carjacking, theft, and burglary, and I like my organs inside my body, thank you very much. But I think witnessing firsthand the stark wealth disparity between El Paso, Texas, and Juárez, as I drove between the border cities, formed one of the first cracks in those starstruck rose-colored glasses I wore so proudly. I thought I was on a path that led to a utopia, but the road was paved with poverty.

The Rio Grande splits what should have been one giant city into two—one in Mexico and the other in the United States of America. As you drive along I-10, north of the river, you can easily compare the extreme lack of infrastructure and blatant effects of economic struggle on one side to the shimmering skyscrapers, white pavement, and wide streets devoid of imperfection on the other. Such foreshadowing, I think, looking back on it. After all, now I question whether we should have borders at all, often considering the ways in which territorial claims-making influences other forms of social, political, and economic disparities and will continue to do so on extraterrestrial colonies.

The individuals who ran the Spaceport America emergency services cautioned me so intensely about crossing the border, but some of the few emergencies they had to handle were from those *running* the border. Undocumented immigrants seeking shelter, an escape, a new life, at a Spaceport selling flights to low Earth orbit that they would never be able to afford. These were moments when individuals fleeing would run the Mexican border and cross paths with Spaceport America—intentionally or not—and the Spaceport emergency services would need to temporarily handle the situation until border control contained the undocumented immigrants.

During my fieldwork, I spent a lot of time with the emergency services and security team and heard many of these stories. The spaceport was legally required to have extensive fire and emergency services due to the fact that they were *launching things into space*, so it was surprisingly well outfitted for a place set in the middle of nowhere. And its remote location—about an hour from the nearest police station—meant that a couple of the Spaceport America fire-fighters were deputized to handle legal emergencies if they should arise. Such emergencies were infrequent and usually the result of drunken exploits from the locals, or issues with undocumented immigrants, but they did happen. At the time, though, major launches from the spaceport—such as Virgin Galactic flights—were infrequent, leaving the firefighters, EMTs, and security guards, well . . . bored out of their minds.

They were downright excited to take a break from cleaning duty and rattlesnake relocation to show an anthropologist around (and yes, I did get to ride in the fire truck).

Fieldwork takes many forms, as I mentioned earlier, but much of it consists of what Clifford Geertz termed "deep play." In other words, you have to spend as much time as possible with the culture you're studying. Be up-front with your positionality, attempt to maintain neutrality, and then do as your participants do. If I had been studying a gang in Juárez, this would have involved participation in quite a bit of illegal activity, and trust me, anthropologists don't shy away from such controversial behavior. For example, Alice Goffman is well known for her research on police mistreatment of impoverished Black communities in Philadelphia. Her work showcased drug abuse and violence and resulted in her being accused of conspiracy to murder.[5] Likewise, Jeff Schonberg and Philippe Bourgois's *Righteous Dopefiend* follows a group of heroin and crack addicts for over a decade as they study a moral economy of sharing and the political and social ramifications of long-term drug addiction in society.[6] Their work also involved their accessory to crime, as well as moments in which they saved the lives of individuals who might have otherwise died from an overdose.

Yet I remained sheltered in the space industry's desert utopia. My fieldwork involved fewer drugs and less police brutality and more sitting in on meetings, attending city council sessions, shadowing the Spaceport America team, visiting their tenants—who were primarily

UP Aerospace and Boeing at the time—and tagging along to happy hours and barbecues. In the evenings, I came back from either Spaceport America headquarters or the spaceport itself to write field notes, jotting down journal entries and poignant moments. Occasionally I would conduct formal interviews, but they were far less useful than the informal conversations that happened over dinner parties or on the drive out to the spaceport.

I befriended firefighters and drank bourbon with Virgin Galactic's employees. Talked about science fiction and Dungeons & Dragons with Boeing's project members. Had coffee with Spaceport America's general counsel and volunteered for elementary schoolers with their director of aerospace operations. I ate pizza and reminisced about raves with an engineer working on the *Starliner* capsule. Visited Elephant Butte Dam with the only employee my age. I got asked out four times. This was what fieldwork looked like.

Twice a week the Spaceport America employees and I would go out to the Spaceport America base together, driving the hour-and-forty-five-minute journey through unpaved desert as a group rather than taking separate cars.[7] The Chihuahuan Desert, which surrounds the spaceport, is a fearsome thing. It's stark and otherworldly. Broken up by hazy mauve and burnt sienna mountain ranges, the landscape could easily be mistaken for the red planet, or a far more fantastical place not yet discovered. With an average precipitation rate of ten inches a year (compared to America's average of thirty-nine) the area is dry and arid, with temperatures often hitting one hundred degrees Fahrenheit or higher. We would begin a trip to the spaceport by meeting at the office in Las Cruces, on time (I swear), but we would always leave an hour late, because coffee mattered and the cars needed to be filled with gas and the windshield freed of yesterday's mud. Dr. Bill Guttman, Director of Aerospace Operations, would drive, and Chris Lopez, Director of Site Operations, would ride shotgun. Most days I would sit in the middle of the back so I could hear everyone clearly while they told me stories of past launches, special guests, strange occurrences, and close encounters with the foreboding landscape around us.

It was because of that landscape that I would occasionally unbuckle and scoot up to the window. I wanted to make it easier for my companions, all born and raised locals, to tell me about the environment.

"Those tumbleweeds are invasive. See how they pile up on the side of the highway?"

"Look down there. In the rainy season the Rio Grande will flow through this canyon."

"Do you see the mountains? Don't they look like their name? Organ Mountains. Can you see the organ pipes?"

"See those . . . they're oryx, not antelopes. The only species in their genus."

*Oryx gazella.*

The highway would only take us so far, and after about an hour and a quick pass through border patrol—an outpost on the American side—we were driving over unpaved dust-coated roads. (The road from Las Cruces to Spaceport America was paved not long after I completed my fieldwork.) For a moment, I thought that surely we were all going to die, because when Guttman drove he swerved with compassion not just for oryx and cows but also for snakes, lizards, rabbits, and roadrunners. If he could have seen the ants from the driver's seat, he would have swerved for them too.

During one drive we slowed near a gorge and looked out at the horizon. Guttman and Lopez saw something I didn't see.

"It'll rain soon," Guttman said.

"I know. I'm really hoping they pave this section before the wet season hits," Lopez added.

"Why?" I piped up.

"If the rains come, the roads will wash away, and even with the four-wheel drive we won't make it past. We will have to detour through T or C."

"Wouldn't that be—" I started to ask.

Lopez answered me quickly: "About two extra hours."

"Why come at all then? You'd either have only one or two hours out here or at least a twelve-hour day."

Lopez laughed, but he was only half-joking when he said, "We have to. The spaceport needs us."

Some of the simplest but most lethal risks that I could have encountered during my fieldwork were environmental: heatstroke, getting lost in the desert, dehydration. When rain occurred it often caused flash floods that would wash away the road to the spaceport. Dust was a constant problem on the horizon, causing eye irritation and poor visibility. While completing my fieldwork, I encountered multiple tarantulas, rattlesnakes, scorpions, antelopes, and cattle (the cause of many car accidents), and happened to sit on a wasp's nest one morning while taking my breakfast on the porch of my Airbnb, landing myself in the emergency room. The desert flora was not any tamer: the thorns of mesquite bushes grew thick enough to pop tires, and invasive tumbleweeds destroyed the overall health of soil and the surrounding landscape.

Guttman and Lopez never left me alone. Undoubtedly because they weren't allowed to, but also because they knew I was clumsy enough to fall into a nest of rattlesnakes.

But in general, the Spaceport America team was as apathetic toward as they were cognizant of these dangers. They faced them with the same optimistic fatalism with which they faced all complications of space exploration.

Sacrifices had to be made for the sake of progress.

Whether these sacrifices took the form of a two-hour reroute, increased taxes, or the death of a "test god."

One day while standing out at the Vertical Launch Area, Lopez turned to me and urgently said, "I harken back to my family's immigrant days [when I think of the space race]. One of my cousins explained to me that someone has to take the leap. In my family's days, there were four sisters that immigrated to the country. The matriarch could have stayed back in her own country and known the path of her life, but she wanted to dream bigger. She wanted to have more possibilities. And now I am very proud of my family, which has accomplished many things for many people and many generations. Investing in the space industry is the same. And if we don't take the leap we'll be left behind."

He, like many space industry employees, feared the end of the human species and saw commercial space ventures as a step toward human colonization of other astronomical bodies, and thus the perpetuation of the human species if all else were to fail on Earth.

In a similar fashion, Karen Barker would often remind me of the sacrifice that the citizens of Truth or Consequences made when they voted to contribute taxpayer dollars to help construct and maintain the spaceport. Words like *faith*, *conviction*, and *promise* were used often by Spaceport America team members. So too were they used by locals and city council members. The spaceport represented something far greater than a tourism industry yet to blossom. It was a source of survival for a town holding its breath in fear of exhaling too hard and watching its remaining population disappear like dust on the wind.

And that fragility extended beyond the town to the future that the spaceport was prophesying. It was clear from the beginning that this was another utopia that gods were building on the backs of those who would never touch heaven.

And that they weren't divining a future that was guaranteed.

Driving through the streets of Truth or Consequences, I could not shake a strange sensation that formed in the pit of my stomach. Spaceport America was supposed to bring new life to Truth or Consequences's economy. Instead, Virgin Galactic and Spaceport America headquarters were based in Las Cruces (a city with a population twenty times that of T or C), and almost everyone commuted from either Las Cruces or Albuquerque. Very few spaceport employees actually lived in Truth or Consequences, and the jobs that the spaceport brought to New Mexico were occupied largely by exported executives from other states.

I didn't know it then, but it was *doubt* weighing down my steps. Doubt and *guilt*. I wanted to believe that human space exploration was the answer but struggled to see any righteousness in "the problem" that commercial space ventures were attempting to solve. Why couldn't we take care of issues like poverty and scarcity on Earth and *then* go colonize other planets?

Was one speculative future truly less achievable than the other? Could humanity perhaps work toward both goals simultaneously?

And what about the Earthlings left behind? What happens to them? Who decides who goes and who stays? Who really gets a voice in the matter? It certainly wasn't the firefighters or the EMS team. It wasn't the bartenders or grocery store workers in Truth or Consequences. It wasn't the undocumented immigrants running the border.

Yet they too were citizens of planet Earth.

News outlets heralded Spaceport America as New Mexico's "hope for a revived economy," but local inhabitants of Truth or Consequences weren't exactly getting factored into flight rosters or profiting from ticket sales.[8] In fact, Spaceport America employees, tenants, and customers started to remind me eerily of colonists. They certainly didn't *feel* like natives. Some *were*, but many weren't, and few engaged heavily with the locals. Even tours of the facilities were (at the time I was there) heavily restricted. Tourists were limited to what was essentially a drive-by tour of the facility, available by reservation only one day a week. Tours of the Virgin Galactic hangar facility and the Spaceport America offices on base were not allowed.

Who was really benefiting from the project and who got a say in why or how humans went to outer space? What's more, were there consequences to human space exploration that weren't being considered?

My participants did not take kindly to these musings.

# 4

# ASTEROIDS AND ACCESS

*Old Mesilla, Las Cruces, NM, 2018* CE

It's worth pausing once more on both the concept of *positionality* and *privilege* before continuing. Specifically, *my* positionality and position of privilege.

I don't think tours of Truth or Consequences or headlines about undocumented immigrants fleeing the Mexican border were the only things heightening my growing interest in wealth disparities and issues of access.

Remember that childhood I told you about? The *privileged* one, filled with loving, supportive parents who thought the world of my education? It moved in stark parallel with a rebellious, destructive, chaotic phase in my youth that I am lucky to have escaped without a criminal record. A time in my youth when I was surrounded by friends who lived below the poverty line. Though my family could have bought me a ticket onto a Virgin Galactic flight if they had wanted to, the people I loved the most would never see those stars.

My South Florida hometown had a middle school geared toward the arts that nurtured and cultivated intelligence and creativity. I thrived in that environment and still maintain contact with some of my middle school teachers. But once I entered ninth grade, I felt trapped in a public high school populated with youth from both the state's wealthiest neighborhoods and some of its most impoverished. I found myself skipping classes that were assigning literature I had read by the third grade. My history teacher pronounced *plague* as *plack* and thought Franklin D. Roosevelt had been the president during World War I. And I wasn't only fighting a poorly organized public school system. I constantly fought with those supportive, loving parents of mine over politics, religion, morals, and beliefs. I struggled to fit in with the popular kids and the sporty kids. Related more to the theater teacher, and then gradually the stoners and "artistic freaks." Most of those stoners and "artistic freaks" didn't come from wealth. Many of them were raised by parents who bought groceries with food stamps and had done so their entire lives. Several grew up in environments shadowed by addiction, crime, and abuse. But several of them also shared my inclination toward eclectic media, music, fashion, art, anything that was abnormal and stimulant, cerebral hobbies, and the avant-garde.

We utilized our free time, attained immorally through truancy, to our advantage, hunting down books by French philosophers in the city library and playing inebriated games of chess by alligator-infested Florida canals. We skanked to ska music and listened to swing and bluegrass and zydeco in the rain. We made poor attempts to dance the Charleston. Collected instruments, broken things, and other lost children. Lit bonfires on the beach and regrettably woke up the next morning with sand in strange crevices. We fell in love with each other hard and often. And during the midnight hours of feverish swampy winter nights, we traced the constellations of Orion and Canis Major across the sky, pinpointing Sirius with our fingertips.

During those years, I felt *almost* entirely accepted by that group of outsiders.

*Almost.*

Between spoken-word poetry performances and half-howled gypsy punk lyrics, snide comments about my own upbringing slipped through the cracks. They questioned whether I would narc on them and whether I would cover their backs when child's play became adult problems. In those days, synthetic drugs abounded, as did psychedelics and, eventually, as my friends aged, cocaine and other more serious drugs. It didn't matter how many times I followed in Clifford Geertz's footsteps and ran with them over fences and through Spanish moss–covered forest to escape whatever mess we had gotten ourselves into. That wealth gap was a yawning, festering wound that even fifteen-year-olds could smell putrefying between us. And I despised the problems the wealth disparities caused, without understanding that the source of their anger, for it was something entirely different from the source of mine. I wanted to belong and hated that my background set me apart from them. They hated that I wore brand-name clothes and grew up with a maid.

Sophomore year I skipped 100 out of 180 days of high school while maintaining As and Bs. When called into a concerned adviser's office, I remember telling her "that clearly, the problem was with the public school system, if I could skip so much school and pass."

She didn't respond and ended the meeting without consequence.

Junior year—without telling anyone—I enrolled in twelve courses instead of the usual six or seven that a traditional high school student takes and graduated a year early. Turns out you don't need parental permission to do so until it comes time to file for the official graduation diploma.

My parents were blindsided by my early graduation, to say the least. But I was done with high school. Running laps during physical education wasn't teaching me anything. Dressing up in uniforms and sitting through fifty-minute blocks of halfheartedly taught courses wasn't teaching me anything. These were things I could learn in my own time.

The memory of a friend's mom loaning her sixteen-year-old her bong because she was pregnant, that taught me something.

Having someone's dad pat you on the back while tears stream down your face and say, "It's OK, hon, everyone runs from the cops a couple times in their life," that taught me something.

Going for "water runs" during a party because the well water had long been declared carcinogenic but the city had not acted on it for over a decade, that taught me something.

It didn't teach me that these things were morally right. It taught me that this world existed, not as better or worse but different . . . and that privilege and access were powerful things.

And so I left and moved on and grew up. I blossomed in college. Absolutely flourished with an educational path designed to my needs and interests. I felt an intense kinship with university faculty across disciplines and took eclectic courses on niche topics I knew nothing about. The history of the Rhine, Arctic peoples, madness and hysteria, multicultural architecture.

I got four degrees in six and a half years and am currently working on my fifth.

Don't think me too accomplished, though. I also got rejection letters from seven out of eight Ivy League schools and have no doubt that if I had ever applied to Dartmouth, they would have rejected me too.

But I did come back to South Florida. Almost every Thanksgiving and Christmas, when I came home to visit my parents for the holidays, I would swing by that neighborhood—the place where I *really* grew up—and spend at least one night hanging out with my friends. Most of them were still there (except one who went off to become an opera singer, which was pretty cool).

Post high school (if they had graduated) these individuals were doing their best to stay grounded here on Earth. To keep a job. To feed a kid. To support an elderly relative. To get a promotion. To stay sober. To get off probation. To pay rent. To stay out of jail. To save up for a house. Our conversations about space and the career I was starting to build were hard to maintain—and not because of a lack of comprehension. My old friends simply had little faith in their own voices on the matter. And that is the point of this interlude.

Society had given them no voice in celestial matters.

And in all honesty, they thought the whole "space thing" was kind of dumb.

"Go to space? Can't I just get some health care, man?"

My annual visits to South Florida, during the years when I was working most intensely for the space industry, were an increasingly stark reminder of how privileged conversations about space exploration were. Not only that, but they forced me to consider other, more worldly concerns. Concerns about scarcity, about education, about housing, and costs of living. Concerns that existed right on the doorsteps of individuals I knew, not on other planets.

---

I wish I could remember where we had coffee that day, during my last week at the Spaceport. I know it was just outside of Las Cruces in Old Mesilla. I remember running my fingers along an adobe stone wall near the town center as I walked toward the coffee shop. Behind me stood the Basilica of San Albino, built just after the Mexican-American War. They say that during frontier times, the town was frequented by bandits like Billy the Kid and Pancho Villa. But memory is a fickle thing, and I have no doubt the space lawyer I spoke to that day remembers our conversation differently than I do, just as I imagine American textbooks remember the Mexican-American War differently than Mexican textbooks.

I certainly remember that weeks later, when I published an academic blog post based on the themes we discussed—mainly space law—she was upset by the way I translated our conversation.[1] Frustrated, I think, by the connections I made, because they weren't blindly optimistic about the future of humans in space. But I'm not the only one to have published articles criticizing the ambiguities of space law and the intentions of commercial space companies. The foundations of space law, such as the outer space treaties and the Commercial Space Launch Act of 1984, were based on the dynamics of the Cold War and competition between two leading nations: Russia and the United States. In the decades after the fall of the Berlin Wall, space law became increasingly complicated by the inclusion of more nations with launch capabilities and space interests and the creation of new space policy with diverse motives such as space tourism and resource extraction.[2] In light of this new multidimensional world of space law, the outer space treaties have been increasingly criticized and called for revision.

But the space lawyer was one of the last individuals I interviewed at Spaceport America, and potentially one of the most helpful. Stars were beginning to form constellations in my mind, and those herculean questions that my fieldwork helped me form started to at least meet their respectable case studies and methods of answer. Who really benefited from human space exploration when access was granted only to those who had the money to get there in the first place? Should humans explore space at all? What cultural outcomes would result from continued human space exploration and the colonization of other worlds? Who got a voice in decisions about space exploration if not, realistically, the common citizen?

Even during those final weeks of research, my work was still very speculative, theoretical, and abstract in nature, but so too were the mission statements of the companies I studied. Those organizations and I were laying necessary foundations for a greater body of research. And this isn't uncommon for an anthropologist. Often one enters the field with research questions and initially gathers data without too much intent. Intent might bias things in one direction or another. Only after fieldwork ends does a researcher piece everything together.

What started between the space lawyer and me as a conversation about space law over a latte would serve as the answer to some of my primary research questions about human space exploration. But it would also lead to a stellar construction of constellations that I couldn't have predicted.

How quickly a trajectory can change! One minute I was thinking about spaceports and the next asteroid mining. Orbital maneuvering of the mind reveals answers hidden behind mental debris. My fieldwork took an unexpected turn in its final week.

I'm not surprised our conversation strayed from spaceports to space law and asteroid mining that day. In November 2015, the United States had only just passed the US Commercial Space Launch Competitiveness Act, which included Title IV on Space Resource Exploration and Utilization. The act is an extension of the 1985 Commercial Space Launch Act. Section 51303 of the new US Commercial Space Launch Competitiveness Act gave any US citizen engaged in commercial recovery of an outer space resource the right to "possess, own, transport, use, and sell the asteroid resource or space resource obtained in accordance with applicable law, including the international obligations of the United States."[3] This led to a wave of increased commercial activity related to asteroid mining and extraterrestrial resource transportation and resulted in the establishment of several start-ups dedicated to asteroid mining, such as Deep Space Industries, Trans Astronautica, and Planetary Resources. It also led to NASA funded asteroid mining research and the creation of a graduate degree in space resources at the Colorado School of Mines.

In the initial years after these corporations were founded, extraterrestrial resource extraction was framed as the next stage in space exploration—a techno-scientific venture to be prioritized, which would have economic impacts both on Earth and for deep space exploration.[4] But only a decade after the passing of the act, these companies had gone bankrupt or reframed their celestial goals. Some experts on space exploration claimed that extraterrestrial resource extraction could not be accomplished due to lack of financial support. Others claimed that the necessary technological infrastructure was not achievable. And others still insisted upon more emotional, political, and cultural reasons for the industry's deterioration.[5]

Before moving forward, though, I think we need to interrogate the *mythology* of the most speculative promises of the asteroid mining industry alongside the very real issues of privilege, access, economics, and politics. And this is where I really start to draw on my training, not just in anthropology but in history. Those childhood obsessions with foreign fables and mythos have come in handy—oddly enough.

As the space lawyer and I discussed the complexities of space law in relation to the rise of the asteroid mining industry, I was reminded of something very specific. A utopia not unlike Spaceport America, also accessible only to a heroic few: Ἠλύσιον πεδίον, Elysium or the Elysian Fields.

Elysium is a specific place within the ancient Greek concept of the afterlife reserved only for heroes and children of the Gods.

Hold the concept of it in your mind for a moment, with as much sincerity and conviction as you would the potential of asteroid mining. Both are speculative in conception. One is a speculation based on an ancient religious belief system and the other a speculation based on an economic belief system. I came away from my conversation with the space lawyer wondering what the cultural consequences of asteroid mining might be. I was already concerned about how accessible space exploration was to the common citizen and who got a voice in space activities and development. What would happen if those who had access to space could now harness valuable space resources? Would this result in a world further divided up into paradise and purgatory, or "Elysium" and . . . everyone else? A world where some voices were included and some were left out and where current wealth disparities increased further than they already had?

By analyzing the history of outer space law and interrogating both its promises and inadequacies we can understand what it might look like when the twin mythologies collide. We can question how the division between classes that exists in the Greek concept of the Elysian Fields and commercial space enterprises comes to be and what role it serves.

But the only way to examine what exactly has been negotiated and promised to the people of Earth through space law is to pass through the gates of Elysium and confront the utopia itself. For that to occur, and for us to learn the destiny of humanity as it has been woven by the Fates, we must first humbly allow our thread of life to be cut.[6] This deadly task was the role of the Fates, or Μοῖραι (the Moirai), as they were called in ancient Greek. These three Goddesses acted as *gatekeepers* to the afterlife—an important part of our tale that we will return to in the next chapter.

I know, I'm speaking in metaphors (again), but as an exercise let's keep these two visions—one of ancient mythology and one of, dare I say, science fiction—in the forefront, as we investigate the rhetoric flowing from the mouths of commercial space organizations.

# 5

# A TRIP DOWN THE RIVER STYX

*The Elysian Fields, Greece, 700 BCE*

**In the late 1960s and throughout the 1970s,** the Fates began to weave a complicated tapestry of treaties and principles related to outer space. These were developed by the United Nations Committee on the Peaceful Uses of Outer Space and were inspired by the principles of the Antarctic Treaty, as well as complex Cold War tensions at the time. It was President Eisenhower, in his address to the United Nations General Assembly in September 1960, who first proposed the creation of an outer space treaty that "would enable future generations to find peaceful and scientific progress, not another fearful dimension to the arms race, as they explore the universe."[1] The urgency in his demand was partly in response to the Soviet Union's rejection of the US proposal to require international verification of the testing of space objects after the recent invention of intercontinental ballistic missiles.[2]

In the early decades of the Apollo space era, Eisenhower was cautiously navigating a response to Soviet space exploration. Though the American space program was as nationalistic as the Soviet space program, Eisenhower was careful to frame space development in a civilian rather than militaristic light.[3] In 1958, six months after the United States orbited its first satellite, Explorer 1, Eisenhower signed into action the creation of the National Aeronautics and Space Administration. The development of NASA as a civilian space agency was part of the effort to frame space in a civilian light. Eisenhower feared that to militarize space exploration would be to project an image of aggressive intent and that this would have repercussions for the Cold War.

Eisenhower's request for an outer space treaty was successful. In 1967, the United Nations Committee on the Peaceful Uses of Outer Space met and developed the first such agreement, formally known as the Treaty on Principles Governing the Activities of States in the Exploration and Use of Outer Space, Including the Moon and Other Celestial Bodies. "The Outer Space Treaty," as it is colloquially referred to, is broken into seventeen articles, which emphasize that the exploration and use of outer space, "shall be carried out for the benefit and in the interests of all countries, irrespective of their degree of economic or scientific development, and shall be the province of all mankind."[4] The treaty also outlined things such as the ownership of outer space, use of nuclear weapons or other weapons of mass destruction in outer space, and avoidance of harmful contamination of space and celestial bodies.[5]

The Outer Space Treaty was the first of five and was shortly followed by the Agreement on the Rescue of Astronauts, the Return of Astronauts and the Return of Objects Launched into Outer Space or "the Rescue Agreement" (1967), Convention on International Liability for Damage Caused by Space Objects or "the Liability Convention" (1972), Convention on Registration of Objects Launched into Outer Space or "the Registration Convention" (1974), and the Agreement Governing the Activities of States on the Moon and Other Celestial Bodies or "the Moon Agreement" (1979).

And in 1984, just before the dissolution of the Soviet Union, Congress passed the Commercial Space Launch Act—the first of four acts that would promote the privatization of space exploration. The Commercial Space Launch Act was designed to harness services offered by private information technology services, remote sensing technology, and telecommunications industries. More significantly, it recognized commercial industry's ability to develop launch vehicles, satellites, and other services.

When considering asteroid mining as a case study, let's focus first on the Outer Space Treaty and the Moon Agreement. Even though it was the Commercial Space Launch Competitiveness Act of 2015 that would eventually legalize extraterrestrial resource acquisition, the Outer Space Treaty and the Moon Agreement were the first to speak specifically toward concepts of ownership and outer space resource acquisition.

Like the Outer Space Treaty, the Moon Agreement makes several declarations regarding peaceful operations and exploration, prohibition of the use of certain weapons, acts of hostility, and use for the benefit and interests of *all countries*, irrespective of their degree of economic or scientific development.[6] The Moon Agreement also adds that "due regard shall be paid to the interests of present and future generations as well as to the need to promote higher standards of living and conditions of economic and social progress and development in accordance with the Charter of the United Nations."[7] In other words, space development should benefit all and the benefits of space exploration should be distributed. So how have these treaties been impacted by not just the legalization of asteroid mining but also other commercial space activities? And what exactly do the Outer Space Treaty and the Moon Agreement promise?

They promise that what occurs in outer space and on the moon will benefit all countries, irrespective of their degree of economic and scientific development. That "outer space, including the moon and other celestial bodies, shall be free for exploration and use by all States without discrimination of any kind, on a basis of equality and in accordance with international law, and there shall be free access to all areas of celestial bodies" and that "there shall be freedom of scientific investigation in outer space, including the moon and other celestial bodies, and States shall facilitate and encourage international co-operation in such investigation."[8] These treaties also promise that the moon and its resources are the "common heritage of mankind," that no one person or government can own the moon or its resources, that anyone has the right to explore the moon, and that all benefits derived from space resources will be shared equitably with special consideration of developing countries.[9] In other words, the Outer Space Treaty and the Moon Agreement seem to describe an Elysium—or utopian frontier—with its doors open to all.

However, the Commercial Space Launch Acts of 1984, 2004, 2015, and 2023 are focused on American goals and the US commercial space economy. The Commercial Space Launch Competitiveness Act of 2015 granted any US citizen engaged in commercial recovery of an outer space resource the right to "possess, own, transport, use, and sell the asteroid resource or space resource obtained in accordance with applicable law, including the international obligations of the United States."[10] Though the Outer Space Treaty says that no nation can have sovereign control over outer space resources *in situ*, the Moon Agreement adds context that once they are *ex situ*—or dug up—then they become the property of whoever dug them up. One space law professor at the University of Nebraska–Lincoln phrased it like so: "Proponents of asteroid mining view the ban similarly to the 'global commons' status of the high seas: no state may colonize the Atlantic Ocean, yet anyone can harvest its fish."[11] Now, not only are Elysium's doors wide open, but outer space resources are ripe for the taking.

The Outer Space Treaty and the Moon Agreement place a heavy emphasis on equity and global benefits, specifically for developing countries, but terms such as *equity, global,* and *developing countries* are suspiciously absent within the commercial space acts. Instead, focus is placed on United States commercial and industrial development and international *competition*. For example, section 115 of the US Commercial Space Launch Competitiveness Act states, "It is the sense of Congress that state involvement, development, ownership, and operation of launch facilities can enable growth of the Nation's commercial suborbital and orbital space endeavors and support both commercial and Government space programs."[12] One has to ask what mechanisms are in place to ensure the distribution of outer space resources to developing nations and to ensure their distribution "equitably" if the US Commercial Space Launch Competitiveness Act essentially implies "finders keepers."

The answer is, no mechanisms—at least none that go beyond ratification. The outer space treaties and the commercial space acts don't work well together. Though the Outer Space Treaty has 110 State Parties, with another 89 countries that have signed it but have not yet completed ratification, there is a lack of enforcement mechanism for its articles and no defined threshold for what constitutes a violation.[13] For example, authors Ishola et al. point out that the Outer Space Treaty did not prevent countries such as China, Russia, France, and the United States during the arms race from embarking on atmospheric nuclear tests.[14]

OK, but hang on, what exactly lies beyond the gates of this metaphorical Elysium? A world divided by those who have access to space resources and those who don't.

The Elysian Fields (sometimes referred to as the Elysian Plains, the Fortunate Isles, or the Isles of the Blessed) were described by Homer as a place "where the living is easiest for mankind. No snow's there, no strong storms, nor ever any rain, but a constant light western breeze coming up off the ocean to provide mankind with cooling relief."[15] Likewise, by Hesiod as a place "untouched by sorrow" and by Plutarch as a peaceful place freed from "tyranny and wars that would never end."[16]

The future that asteroid mining will bring with it was described by companies such as Trans Astronautica, Planetary Resources, and Deep Space Industries as a development crucial for interplanetary space travel and continued advancement of industries here on Earth.[17] For example, mined $H_2O$ could be used for in-space refueling and rare Earth metals can be sold back on Earth. But I speak in the past tense when referring to these companies because, like I mentioned previously, several have either been acquired by other commercial space start-ups or remarketed themselves to fit a new commercial space industry niche. The promises of Elysium have broken. So, let's now cross back over the river Styx and face those who have been left behind.

───────────

The ancient Greek Elysian Fields were much like Spaceport America in that they are certainly *not* open to all. Only heroes and the children of the Gods and Goddesses spent their afterlife in the Elysian Fields. For example, *The Odyssey* speaks of "fair-haired Rhadamanthus," a King of Crete and demigod, dwelling in the Elysian Fields.[18]

It might be argued that the Outer Space Treaty and the Moon Agreement give any government the right to go to outer space, and likewise the US Commercial Space Launch Competitiveness Act extends that past the private sector and into the public sector, but who really benefits from space exploration when access is granted only to those who have the money to get there in the first place? Who gets to enter Elysium in our modern mythology?

Space exploration is not cheap, and space travel requires launch capabilities and spacecraft to leave Earth's orbit. Our problem is one of *equality* versus *equity*. Sure, the Outer Space Treaty, the Moon Agreement, and the commercial space acts leave Elysium's doors open to all, but *only* if one has the money to pay the ferryman and cross the River Styx in the first place.

This was another obstacle the ancient Greeks placed in the paths of the dead. Souls must bring with them a bribe for Charon, the ferryman, in the form of an *obol*—sometimes referred to in Latin as a *viaticum*—or a coin. Before the dead even reached the gates of the Elysian Fields, if they did not have a bribe, they were weeded out.

This was not unlike modern-day attempts to break out of cycles of poverty. What I'm arguing, through this unusual allegorical mingling, is that outer space resource extraction might be framed as a prosperous frontier open to all. But rather than result in wealth, I believe that outer space resource extraction may result in further wealth disparities and an extension of the cycle of poverty into outer space.

Accessibility does not equal equity, and there are no standing regulatory mechanisms to ensure that developing countries benefit from the Outer Space Treaty and the Moon Agreement. Not only that, but the treaties are outdated and lacking a substantial body of signatories in some cases.[19] Nor are there any existing support mechanisms to help developing nations get to space in the first place. Unless regulatory mechanisms are developed to enforce distribution of outer space resources as they are mined, the mining of outer space resources will result in increased wealth disparities and an extension of the cycle of poverty into outer space. An *Elysium Effect*, so to speak.[20] Wealthy, privileged nations or commercial space companies based out of wealthy, privileged nations will only gain further wealth via the mining of outer space resources, and the gates of Elysium will remain tightly shut to the rest of the human population.

---

I've never met a mythology or a religion that spoke more fondly of the Land of the Living than the afterlife. Religious texts don't tend to speak highly of mortals, let's be honest. And feel free to prove me wrong! Please. I really dug around for this one. But as an example, consider Virgil's *Aeneid* (to continue along our use of ancient Greek texts) which at one point describes humans and the cycle of life as "clogged of harmful bodies," "but frames born to die," "wretches," and "plagues of the body."[21]

Perhaps that is why texts that speak of life after death tend to be so full of promise, and why during that time at Spaceport America, my participants' narratives about outer space were founded on hope, no matter how hopeless their future might have seemed at the time.

Authors have argued that the US Commercial Space Launch Competitiveness Act has opened the doors for more participants in the new space race, claiming that "space exploration is no longer a domain reserved for wealthy nations alone."[22] But the "participants" mentioned are often already millionaires if not billionaires, such as Paul Allen, Elon Musk, Jeff Bezos, and Richard Branson, as well as the owners of other commercial space companies. Too many voices are getting left out of conversations about whether or not the human race should be going to outer space and how the extraterrestrial environment is being managed. Who are the Fates—or the gatekeepers—in our modern mythology about outer space resource mining? Because as it stands, it is *a mythology*. A volatile daydream held by a handful of start-ups now being intensely critiqued as "untenable" and "insolvent."[23] The Fates are the small, privileged group who get a say in not only who goes to outer space but also how we go to outer space and how we interact with outer space.

I stumbled out of that café in Old Mesilla and back into the city at the fringe of a desert, uneasy at how my conversation with the space lawyer had evolved. A nagging feeling of doubt bloomed like a malignant growth inside me rather than any botanical beauty I could recognize. But one has to remember that acceptance is the first stage of grief and grief is a part of death, and our trip to the Elysian Fields started with allowing the Fates to end our lives. My rose-colored glasses had not yet shattered, but it was during that interview that I first recognized that I was *wearing* them. My original research questions going into my fieldwork at Spaceport America had been about why people *wanted* to go to outer space. I wanted to examine cultural motivations for human space exploration. But it wasn't until my very last week in New Mexico that I began to ask why everyone was so OK with leaving the rest of the human population behind. I started to wonder not why humans wanted to go to outer space but *what would happen* when we went? What were the repercussions of going to outer space?

Accepting that the consequences of the Elysium Effect was an eminently real part of outer space exploration was the first step to unraveling fate's tapestry in its entirety.

# 6

# SIX DECADES OF SPACE PROTESTS

*Kennedy Space Center, Cape Canaveral, FL, 1969* CE

**O**n the day of the protest, I walked up to the second floor of St. Pancras Station, sat down, and stared out the window at the mass of people holding signs, marching the streets of London.

It started to rain.

My field notes hung limp in my hands.

Ten minutes went by.

I considered ordering a coffee.

Was it strange that the cafés were open at a time like this?

Was it equally terrible that I wanted to play the part of scholarly voyeur, coffee in hand, rather than participate in the protest itself?

Perhaps.

But the sheer force of this demonstration brought to mind an old anthropological work by Nobel laureate Elias Canneti: *Masse und Macht* ("Crowds and Power").[1] In incredibly poetic form, Canneti writes about political philosophy through crowd pathology. And if his work reminded me of anything in that moment it was that to join such a congregation would require absolute surrender on my part. Faith in the cause. Or at least belief and understanding. Lest I be swept away, metaphorically or physically.

I knew that the protest happening outside St. Pancras that day was not *my* protest, but it was an important one. An Earthly one. Alluvial. Carnal. Corporeal.

It was no fated parable. No mythology. It was not a capitalist's prophecy.

*It was happening now.*

I also knew I would most likely be swept away if I joined it.

But my return to London in the midst of such political discord ruptured my starry-eyed innocence. Don't mistake me, I was not rid of my astrophilic desires entirely, but I had acquired from my field-work a deep skepticism and an intense need to question everything regarding human space exploration. No longer could I stand passively by while I watched rocket launches or listened to speeches by space entrepreneurs. And I don't say that with any sense of *revolutionary* pride. At the time a large part of me still *desperately* wanted to forget what I had seen and heard and submit myself to any celestial deity that would take me.

Helios, Nyx, Selene.

Itzamná, Awilix, Kinich Ahau.

Nótt, Máni, Sól.

Ra, Khonshu, Nebthet.

Gods and Goddesses of the night, dawn, sun, stars, and sky.

The prospect of protest, of rebellion, of nonconformity, horrified me.

But I had seen beyond the gates of Elysium and discovered the skeletons hiding in the closets of those who advocated for space tourism. I had spoken to those who would never have a say in whether or not humans went to space and those whose land companies launched from. The more I pieced it all together, the more I realized I didn't like what I was seeing. And there was no recovering my naïveté. Now mission statements that leaned heavily on colonial visions and futures and imagined an Earth left behind triggered investigative instincts I earned through anthropological training and firsthand experience.

And I could not stop asking questions.

Not just about who got a say in why or how humans went to outer space or the consequences of human space exploration. But also questions about political process, economics, funding, and what an alternative future to human space exploration could look like.

But my unfaithfulness to what felt like a globally unanimous ambition was isolating. Publishing my first version of "The Elysium Effect" in the *Geek Anthropologist* after I returned from Spaceport America made me want to puke. And when I was invited to write "Lunar Imperialism (and How to Avoid It)" almost exactly a year later, I still pulled my literary punches. I was absolutely terrified that critiquing the only industry I was building credibility in would result in the end of my career prospects, and I resisted the urge to critique certain aspects of it.[2] I was twenty-two and scared of my own voice, and I was still coming to understand the power an anthropologist (let alone a young woman) could have. On top of that, I had yet to encounter much of the literature, or other individuals, who were, like me, laying the foundations for both a cosmic rebellion and an Earthly ascension.

But as hard as it was to imagine protesting Artemis-era ventures, or something like the Apollo missions, such protests did and do occur. And the entwined histories of the two play an important role in any modern-day reckoning.

———

On July 15, 1969, on the eve of the Apollo 11 launch, Reverend Ralph Abernathy led a protest just outside Kennedy Space Center. Alongside twenty-five African American families and four scruffy mules pulling wagons, he demonstrated against the United States' "distorted sense of national priorities."[3] Abernathy and his fellow protesters held signs that made proclamations such as HUNGRY KIDS CANNOT EAT MOON ROCKS and MOONSHOTS BREED MALNUTRITION. His peaceful

Apollo-era protest made three requests of NASA: that ten families of his group be allowed to view the launch, that NASA "support the movement to combat the nation's poverty, hunger and other social problems," and that NASA's scientists work "to tackle the problem of hunger."[4] The four scruffy mules, unyielding and emblematic of manual labor, symbolized the issues Abernathy and his fellow protestors wanted to prioritize: food insecurity, poverty, lack of education, and rising unemployment rates.

Shortly after, in 1970, after the launch of several Apollo missions, jazz musician, poet, and activist Gil Scott-Heron released his spoken-word poem "Whitey on the Moon." One section of it proclaims:

> I can't pay no doctor bill.
> (but Whitey's on the moon)
> Ten years from now I'll be payin' still.
> (while Whitey's on the moon)

The poem speaks out against increased taxes, as well as the same socioeconomic problems Abernathy protested.[5] Meanwhile, headlines such as MOON PROBE LAUDABLE—BUT BLACKS NEED HELP and BLACKS AND APOLLO: MOST COULDN'T HAVE CARED LESS, made the news.[6] Though often unrecorded in the Apollo-era history books, backlash against the moon landings was a salient part of current events. Black Panther Party leader Emory Douglas went as far as to compare the Apollo rockets to a modern-day slave ship in the extremely well-circulated *Black Panther Community News Service*.[7] Douglas's criticism commented on legacies of colonialism, poverty, dehumanization, slavery, and racism, highlighting the imperialist intent of space ventures at the time.

One *Times* article from 1963 stated, "The dollar price is beginning to unnerve some congressmen, and others, with the scientists believe it immoral to put a project of this size, by far the greatest ever undertaken by man, on the level of a teenager's car race."[8] That same article pointed out that though "chances of losing the moon race and the attendant prestige are great, that prestige would soon come to wear thin and that it's likely to be captured in the long run by the side which takes the time and profits from the disasters of the other side." Congressman Chester Earl Holifield was also quoted at the time arguing that the money used on the Apollo program would be better spent on schools, housing, and hospitals.

In 1964, just a year later, sociologist Amitai Etzioni published the book *The Moon-Doggle: Domestic and International Implications of the Space Race*, which provides comprehensive evidence of scientists who opposed the Apollo program. Etzioni argues that scientists challenged the poor distribution of scientific funding, asserting that money should be distributed across defense and Earthly funding and that celestial exploration be left to the robots.[9]

Then, in 1971, high school student and Navajo poet Francis Becenti reacted to human space exploration with his own poetic rebellion, a section of which famously states:

> *So I guess they are going into damn*
> *space again soon.*
>
> *A two and a half billion dollar*
> *camping trip.*

Becenti's poem—an Apollo-era ghost—circulated Twitter in the months after I returned from Spaceport America. Heron's spoken-word poetry was also reanimated in the 2018 dramatic retelling of the Apollo 11 moon landing *First Man*. The film's coverage of Apollo-era protests provoked audiences enough to result in a string of media responses just after its release.[10]

Becenti was not the only Navajo citizen involved in the Apollo-era rebellion, either. During the first years of the Apollo program, NASA leased Navajo land for testing the moon walker. During the testing, NASA invited Navajo tribal chairman Peter MacDonald to the testing site to observe procedures, and he brought with him a Navajo singer and medicine man. The medicine man requested to record a message for the "moon people" in the language of Navajo peoples. NASA obliged, bringing him a recorder. After the message was recorded, they asked MacDonald to translate. MacDonald revealed that the message was a warning to future moon people "to watch you guys carefully because you might screw things up on the moon the way you have on Earth."[11]

Though the 1960s and 1970s were a time dedicated to space exploration and "scientific progress," that does not mean that such ventures were endorsed by all or that "progress" meant progress for the masses. Soviet-era space exploration was notoriously motivated by nationalistic, militaristic desires to win the Cold War, and when Soviet space exploration was framed as technological development "meant for the masses" it was as a way of promoting a new Soviet socialism.[12] Even the famous 1975 "handshake in space"—when Apollo astronauts and Soyuz cosmonauts docked their mechanically dissimilar spacecraft together, supposedly as a diplomatic and techno-scientific effort—was a careful spectacle of political theater, played out amongst a backdrop of stellar proportions.[13] Though to many, 1975 may have appeared as a year of celestial diplomacy and détente, Cold War space age political restlessness was replaced with a social restlessness reflective of issues that stemmed from the Vietnam War, civil rights movement, counterculture movements, and long history of neoliberal concerns that persisted with resilient ferocity.

In other words, the history of protesting human space exploration does not begin with *Ground Control*. Its legacy is as old as the practice of human space exploration itself. And resistance to space exploration—whether in the form of protest or poetry—hasn't gone away. It has simply remained a frighteningly *untechnocratic* secret, reluctantly publicized unless featured in box-office dramas. Because to many (like those at Spaceport America), space exploration equates to *progress*.

In the same way that Luddism—with its purist intentions of technological rejection—feels like a betrayal of advancement, so too does a rejection of human space exploration feel like treasonous de-evolution.

Well, this book is going to explain why that's wrong.

Eventually.

We're getting there.

———————

Artemis-era protests of human space exploration haven't looked that different from Apollo-era protests. Like I mentioned before, some of the same rebellious poetry continues to circulate on social media. In addition, arguments against human space exploration retain similarities to those made in the twentieth century. It is also largely the same communities that have continued to protest human space exploration: minorities, people of color, Indigenous peoples, the impoverished. Occasionally, a celebrity or politician might push back against Elon Musk's or Jeff Bezos's latest space capitalist goals with a spicy tweet,[14] but for the most part, the wealthy do not protest space. They don't have to. They are already the sons and daughters of Artemis and Apollo, and as children of deities can easily enter the gates of Elysium once their comfortable Earthly lives have ended.* Or, to put things plainly—they already have a voice in the matter. A ticket into space. A seat at the table.

Sometimes protests against human space exploration are grounded in Earthly debates about place and space, boundaries and ownership, not on extraterrestrial bodies but here on our home planet and often on Indigenous land. Consider the Mauna Kea protests, which have

———————

* Though according to Greek mythology, Artemis technically never had children.

received increasing publicity due to the Thirty Meter Telescope controversy but have been ongoing since the 1960s.[15] Kealoha Pisciotta, the president of Mauna Kea Anaina Hou, explained that telescope projects on the island of Hawai'i have a history of "going beyond their initial purview, at least in the eyes of the island's inhabitants."[16] Initially in the 1960s, residents of the island were told that only one telescope would be built on Mauna Kea, but within a few years several other telescopes and astronomical observatories were built by organizations such as NASA, the University of Arizona, and the US Air Force.[17] "That broke trust," Pisciotta told *Forbes* magazine in an interview.[18]

Today, Mauna Kea protests continue, as Indigenous Hawaiians fight to preserve land that they consider sacred and culturally significant. Recently, after several years of protesting the construction of the Thirty Meter Telescope, Hawaiian activists won the right to be included in scientific decision making related to Mauna Kea.[19] Now, a new law takes into consideration multigenerational impact, culture, and environmental factors giving Native Hawaiian cultural experts voting seats on a new governing body.

But the right to have a voice in decisions that were made about Indigenous land, and its scientific and cultural value, had to be *won*. This was never considered inherent by policymakers.

---

On April 4, 2017, not long after the Hawaii Supreme Court saw a slew of appeals submitted against the construction of the Thirty Meter Telescope, ten thousand protestors gathered outside the Guiana Space Center in Kourou, French Guiana. The Guiana Space Center, colloquially called "Europe's Spaceport," hosts to the European Space Agency (ESA), the European Union Agency for the Space Programme (EUSPA), the Centre National d'Études Spatiales (CNES or the French Space Agency), the Space Agency of the Republic of Azerbaijan (Azercosmos), and Arianespace.

It is also another highly contested utopian technocracy situated on the coast of an impoverished, postcolonial nation.

Anthropologist Peter Redfield, who I mentioned in a previous chapter, conducted research on the Guiana Space Center for several years, examining its spaceport in the same way I examined Spaceport America.[20] Recent protests continue to reframe his work, such as the demonstrations of April 2017 that brought attention to issues of unemployment, high crime rates, food insecurity, and lack of infrastructure. During the protests, activists blocked launches at the center, in turn threatening the entire European space program.[21] One article quoted protester Bertrand Razan, who explained, "It's a number one global industry, and just next to it, in Grand-Santi, in Sinnamary [nearby towns], there's no potable water, there's no electricity." Another protester stated, "There are rails to transport satellites onto a rocket, and there's no train, no metro, no night bus."[22] Like Spaceport America, the Guiana Space Center is built next to a low-income population who are unlikely to take advantage of the spaceport's amenities.

Kimberley McKinson, another anthropologist, asks similar questions about human space exploration, class, race, and postcolonialism from the perspective of Jamaican postcolonial studies. McKinson asks whether "Black lives matter in outer space." She reflects on space industry histories of racism and how current police brutality issues might teach space ethicists about planetary insecurity and inhospitableness.[23] In other words, what might our present society look like if reflected into outer space, and is that a reflection we actually want to see? What values are we extending into outer space?

Likewise, Chanda Prescod-Weinstein's *The Disordered Cosmos: A Journey into Dark Matter, Spacetime, and Dreams Deferred*, also focuses on systemic racism and sexism in the physics community and specifically how it "crushes the dreams of those who do not fit the traditional mold (i.e. white and male) of scientific 'genius.'" At one point, Prescod-Weinstein specifically argues that "scientists are acting unscientifically when they do not acknowledge the history, philosophy, and sociology of their fields."[24]

We can extend these conversations about protest, activism, and minority communities into conversations about disability. Scholars of disability studies such as Sheri Wells-Jenson, author of the well-known *Scientific American* article "The Case for Disabled Astronauts," and Alice Wong, founder of the Disability Visibility Project, have argued extensively for the inclusion of voices often left out of conversations on human space exploration.[25] They, along with organizations such as AstroAccess and the JustSpace Alliance, point out that individuals with disabilities might not just be suited for space but *better* suited. For example, an astronaut with an ostomy bag could easily avoid the troublesome technical challenges that stem from microgravity bathroom breaks.

In 2022 the ESA announced a parastronaut feasibility project, which requested applications from those who were "psychologically, cognitively, technically and professionally qualified to be an astronaut, but have a physical disability that would normally prevent them from being selected due to the requirements imposed by the use of current space hardware." The project aimed to assess the *feasibility* of the selection of astronauts with physical disabilities. But the ESA's progressive selection process has been criticized as falling short of its full potential.[26] After all, feasibility of disabled astronauts has long since been proven by AstroAccess, which sends individuals with disabilities on parabolic flights on a regular basis. While on these parabolic flights, individuals perform microgravity experiments, test novel safety systems, and conduct other environmental research. Critiques of the ESA feasibility project refer to limitations on the call's "qualifying" disabilities. Some disability scholars take pro-human space exploration stances (while maintaining arguments for inclusion); others share a more decolonial or Earth-focused stance, but all take issues with current imperialist-, ableist-, and capitalist-driven models of stellar exploration.

A year and a half after the protest in London, I was asked to speak at an IMAX showing of *Interstellar* at the National Air and Space Museum by the podcast *Ten Movies*. It was perhaps one of my favorite appearances I have ever done, and certainly an invitation I was incredibly proud of. But when I look back on it in relation to the history of protest I am presenting here, I'm reminded of another poem. A poem by Dylan Thomas featured in the film. My favorite section of which is the following:

> *Wild men who caught and sang the sun in flight,*
> *And learn, too late, they grieved it on its way,*
> *Do not go gentle into that good night.*[27]

Scholarly readings of the poem interpret it as a commentary on death and grief. But when thought of in the context of human space exploration (easy to do, with *Interstellar* laying the foundation for us) I can't help but be struck by the thought that astronauts *do not go gently*. Caught in the flaming exhaust of space shuttles and rockets are cries of protest, unheard voices, and Earthly issues left behind. Not only was I coming of age as an anthropologist at a time when space anthropology was on the rise, but I was also coming of age at a time when a new wave of space ethics was on the rise.

My call to arms, as it turns out, was not so isolated.

# PART II

# A LETTER TO THOSE LOST IN SPACE

# 7

# ON SUCCESS AND FAILURE

*On Board the Concorde, Cruising Altitude, 1976* CE

**I get asked about the future** of space exploration a lot.

About the next hundred years.

Or fifty.

About which commercial space companies will succeed and which will get swallowed up by either their own hubris or bigger space companies or a stilting transformation of goals and values.

I get asked what ideas are sustainable. And which are profitable.

Sometimes the individuals that ask me these questions work for the space industry. But most times they don't. It's far more likely that the person asking me these questions has something more in common with the barista at the café I stopped by on my way to work. Or the bouncer who worked drag nights at my favorite pub. Maybe an undergrad, seeking more than I gave in a lecture. Or an audience member at a conference, looking for anthropological answers to prophetic questions.

And that's it right there. Now we can return to the myth.

People want prophecy. We crave it. Global cultural belief systems are dripping with various forms of foretelling, foreshadowing, the arcane, and clairvoyant. There is an intense, discontent, *vampiric* craving at the foundation of celestial exploration, which stems from dissatisfaction with current reality. Here and now is not enough. We must always look forward and seek answers from the stars.

But I mean . . . jeez . . . who am I to talk? In 2018 I was also seeking answers.

Freshly graduated from University College London, I had four months left on my visa, a master's degree in an incredibly niche field, and no clue what to do with myself.

The possibility of staying in London after my visa ran out was slim to none. It was incredibly difficult to find a job that would sponsor a work permit for an American (trust me, I tried), and the United Kingdom wasn't exactly a mecca for space exploration (though there are spaceports in Cornwall, in Wales, and across Scotland). I knew I wanted to get a doctorate and stay in academia in the long run, but the only program that accepted me at that point wasn't funded. I could join the military. The military was ironically one of the biggest employers of anthropologists. Perhaps because anthropologists understand not *what* we are fighting for but *who* we are fighting.

Either way, the topics of success and failure were definitely on my mind. Both in regard to my own career and the industry I had begun working with. In 2021 Virgin Galactic, the company I worked with so closely out at Spaceport America, finally had a string of successes. They completed their first fully crewed spaceflight that summer— sending their founder, Richard Branson, to space—and by early 2022 they had relaunched reservations at the price point of $450,000.[1] Only a week later than Virgin Galactic, Blue Origin successfully launched a New Shepard spacecraft with their founder Jeff Bezos and his half brother on board. The company has since flown several other crewed flights with high-profile, wealthy passengers such as *Star Trek* actor William Shatner and professional football player Michael Strahan. But these successes were not all sustained. Only months prior to the publication of this book, Virgin Orbit filed for bankruptcy, causing Virgin Galactic's stock to tank just two years after their year of celestial stardom. The potential of space tourism and its ability to take off as a sustainable industry remained in question.

So too, did my own sense of self. There were no guidelines for how to become a space anthropologist, and I was the black sheep of both anthropology programs and space industry ventures. Social science research on outer space was still too new, and too methodologically cutting-edge. It was a very modern and (sometimes) speculative form of anthropology.

So I did what many graduate students did after they earned their degrees: I had an existential crisis. And I worked as a bartender at a pub near Caledonian Road Station.

It was not my first time bartending. I've worked in restaurants and bars since I was fifteen. As I write this book, I'm employed as a bartender at a brewery located not far from Virginia Tech. And oddly enough, working in the service industry has taught me a lot about outer space. From serving tables and cutting off drunkards, I have learned many lessons about team building and human interaction under high stress. High-volume restaurants—as it turns out—make for an excellent extraterrestrial analog site. And I find the service industry work oddly therapeutic, not unlike archaeological digging.

But my existential crisis in 2018 is of relevance because it led to one of those moments of cultural foreshadowing related to space exploration—one based not on the future but on the past.

A prophecy that began with *me* playing the part of the drunkard that needed cutting off.

---

I was regrettably deep into a bottle of whiskey, absorbed in conversation with a lover on a rooftop in central London. The lover was really to blame for what was about to be a burst of inspiration. And don't misunderstand me: good lovers and muses—in whatever form such creatures take in our lives—should be inspiring, as a rule. The inspiration simply came as a surprise because when I was around him, I did my best not to think about space at all, and for the most part, he never asked about it. Instead, we talked about Earthly things. Human things. The struggles and delights and passions that root our species so intensely to this planet.

But that night, he leaned back, cigarette in hand, against a crumbling chimney stack, and looked at me before asking what many had asked before: "Do you think they'll actually make it?"

"Who will make it?" I sat across from him, one leg dangling off the roof. Three stories below us, a group of young adults, dressed for a night out, crossed the road en masse. My companion lived above a pub, and sometimes when he was out of milk or tea, he would slip through a ceiling tile that connected the two spaces and pilfer groceries from their refrigerator. For a brief second, I thought the "who" in question was perhaps the group crossing the road and wondered if his question truly was existential. *Would they make it? In life? Through the night?* London was a city meant for people watching, after all. It was not so uncommon to waste a day in a bookshop window wondering about the lives of strangers crossing the street.

He nudged me with his boot. "The space people. The ones you study."

I sighed heavily and wrinkled my nose as I pulled back from the edge of the roof. "How am I supposed to know?"

"Oh, come on, you just spent months with these people. Do you think there will really be colonies on Mars in fifty years? Or that spaceships will be the next planes?" His question was as impassioned as it was critical. A part of him wanted such celestial imaginings to become a reality.

Above us, light pollution shielded any stars from view. A stark contrast from rural New Mexico, where the Milky Way hung ever-present and crystalline in the night sky.

"Space is the last thing I want to think about right now."

How many job applications had I submitted that week? How

many rejection letters had I received? With a groan, I grabbed the bottle of whiskey and held it tightly against my chest. A mist hung over us like stagnant rain. I had a sense that it was about to start pouring at any minute.

"I think I should return my degree."

He laughed.

"Give up this provincial life of academia! Quit being an anthropologist altogether!"

My words fell flat like apocryphal rain. Halfhearted and heavy.

"Hmpf. I don't think that's possible, is it? You're already an anthropologist. You can't just stop being one, right?"

I paused.

"Well . . . sure I can . . ."

I wasn't sure I could.

"Can you stop being something once you've already become it?"

We stared at each other.

He took the bottle of whiskey from my hand. "OK, but what about after Mars? Will there be starships and—"

"Wait . . ." I paused, remembering what he had said a moment prior. ". . . space planes?"

"Space planes?" he echoed.

"You said 'spaceships will be the next planes.' What do you mean?"

He nodded enthusiastically. "Yeah. You know, like Virgin Galactic—"

I shook my head. "Oh . . . no . . . Virgin Galactic is just short suborbital flights into—"

"—or the Concorde."

"The Concorde? It's not the same. Virgin's spaceships, they . . . they . . . go up and come back down." I mimicked the movement of their parabolic flight path with my hand, moving my arm up in an arc and bringing my hand back down. "The Concorde was a supersonic jet."

"A space plane."

To give him credit, space planes do exist and are vehicles that can glide like a plane through Earth's atmosphere and can also travel into orbit. Sierra Nevada's Dream Chaser is an example, and not long after our conversation, Virgin Galactic partnered with NASA and the Spaceship Company in an effort to develop point-to-point transportation across the globe, not unlike what my friend was imagining.[2] Boeing's X-37 is another space plane that's been successfully launched, and the Space Shuttles are also a type of space plane. Really anything reusable that can go into orbit but come back into the Earth's atmosphere in a similar fashion to a plane is a space plane. It wasn't the concept of a space plane that made me hesitate. No, it was the comparison between the Concorde and the Virgin Galactic spaceships.

I hadn't thought to make it before.

How many times had I been asked to look forward when I should have been looking back? Foretelling is meaningless in isolation; it needs to be contextualized historically, culturally, politically, and *cyclically*. Patterns need to be revealed. I had already begun to do so, in an ancient context, so why not an industrial one? To ask questions about the repercussions of human space exploration and its viability, I should have been looking not just at the Concorde but at other forms of transportation innovation throughout human history as well. For example, the success of railways in Europe but their failed expansion in America. Or other projects such as France's failed Aramis, which is almost identical in concept to Virgin Hyperloop but fell apart after three decades of development. If economists could examine the past to predict economic patterns in the future, why couldn't an anthropologist? If anything, economic patterns were simply one aspect of greater cultural cycles. Answering a question as complex as whether or not it's worth it to send humans to space—and to argue that our species should wait to explore the celestial—requires us not just to look forward, but also to look back. We have to seek out patterns in the way the stardust has fallen across human history.

If we're considering these transportation developments, one could argue that the Concorde was successful, and this is what makes those words—*success* and *failure*—incredibly annoying. Their boundaries and limitations are very difficult to define. Just as space anthropologists have pointed out that the boundary of the "extreme" and the edge of the cosmos becomes more and more difficult to define as human beings continue to breach it.[3] In hindsight, I should never have used such words in relation to my own career progression (and I highly recommend you throw them to the wind as well).

Different models of the Concorde flew for over forty years. The supersonic jet's first successful crewed flight launched in 1969, just months before Neil Armstrong and Buzz Aldrin stepped foot on the moon. Flights were impressively fast, and the Concorde broke the sound barrier for the first time in 1970. The jet reached record speeds, crossing the Atlantic in just under three hours. But such achievements came with complications. The Concorde was noisy, prompting complaints from residents at the surrounding airports it launched from. And not unlike Virgin Galactic flights, flights on board the Concorde were far above the average fliers' pay grade. They also catered primarily to the privileged. A flight from London to New York operated by British Airways cost $7,574 (about $15,000 today) in the 1990s. Airlines who flew the Concorde sold tickets primarily to a wealthy, upper-class demographic. One British Airways spokesperson, Bill Stevens, stated in an article that celebrated the ten-year anniversary of the Concorde, "There is a certain aura about it, a prestige, there's no doubt about it."[4] Stevens described Concorde's passengers as "pop stars, company chairmen, lots of financial institution people, people in a hurry, people who don't like to be in the air too long."[5] But privileged demographic or not, over the years the Concorde struggled to fill its seats, fluctuating ticket prices in an attempt to maintain buyers.

The Concorde also experienced a host of malfunctions and a major crash, which resulted in over a hundred fatalities. This led to the supersonic jet's eventual loss of its Airworthiness Certificate in 2000. Though the Airworthiness Certificate was eventually returned after the completion of an accident investigation, both British Airways and Air France—the two major airlines to fly the Concorde—announced plans to retire the plane in 2003.

I can't help but see the similarities between the Concorde and twenty-first-century space tourism companies when I consider the two side by side.

Virgin Galactic stock shares have declined heavily since their debut on the market in Fall 2019. And a year after Branson flew to space, journalists critiqued Virgin's premier event of 2022: *yet another* vacation retreat for ticket holders who have been waiting decades to actually fly to space. One Twitter follower responded to Virgin's announcement about the retreat by stating, "In all seriousness, I think everyone's a little battle tired from defending VG over the years 'they will deliver trust me' only to read flight delays for another 2 years. To say I am disappointed as an investor is an understatement. I won't be buying into this company anymore."[6] After its rockets failed to launch in January 2023 and the company faced increased financial insecurity, like I mentioned prior, Virgin Orbit—Virgin Galactic's sister company—also lost favor and filed for bankruptcy in April 2023.

Blue Origin faces different criticisms as a commercial space company after a recent launch failure of an uncrewed New Shepard rocket, not from journalists but from Congress. In 2015 a New Shepard booster crashed on landing after an "anomaly" occurred. And after a second launch failure in 2022, which led to the FAA "grounding" Blue Origin pending further investigation, congressional leaders of the House Subcommittee on Space and Aeronautics requested further transparency about the incident. The FAA and Blue Origin released no details about the investigation until January of 2024. Both failures caused the public to scrutinize the regulation and standardization of the development of Blue Origin's technology. Historically, issues regarding standards, transparency, and regulation have been at the forefront of concerns through the rise of the commercial space industry. As private companies gain more and more control over space exploration, how will safety and standards be regulated?

The Concorde is also not the only historic transportation technology that compares to modern-day ventures. Virgin Hyperloop is in development and, as I mentioned, is pretty much the exact same concept as the former French-designed train system Aramis. Both transportation systems were designed to travel as fast as trains but behave more like personalized metros. Neither act on a schedule, and instead are flexible and individualized. Imagine subway systems structured around small pods, about the size of a car, that can be called to specific locations at will. It sounds promising, and Virgin Hyperloop was recently featured at the "FUTURES" exhibition in the Arts and Industries Building on the Washington Mall as a figurehead for "efficient and renewable" transportation technology.[7] Except, after thirty years of development (beginning in 1967 and surpassing prototype stage) Aramis—Virgin Hyperloop's historic twin—caved in on itself, dissolved by the weight of bureaucracy, lack of infrastructure, and deprioritization.[8] Why would Virgin Hyperloop have a different future than Aramis? What gives us faith that technologies that have failed in the past will succeed in the present? What determines technological success in the commercial space industry? Might humanity guarantee the success of human space exploration by breaking certain negative feedback cycles?

Confused by that last question?

Well, here's where I remind you that I *do* want to go to outer space. I don't believe the human species should remain landlocked for all eternity. The question is how to go to outer space without bringing wealth disparities and issues of class, privilege, and colonialism with us.

Ultimately, *prioritization* is the key word when it comes to answering questions about the present and future of human space exploration. So far, I've mentioned "promises" and "prophecies" a lot, and prioritization is how promises and prophecies become reality. One of the major reasons Aramis failed as an infrastructure project was because it was deprioritized by the French government and passed up for a different transportation system, VAL. But when it comes down to it, prioritization is an ordering of cultural and political values and beliefs. So to use prioritization as a tool in future-telling or prophecy-making about an industry, we have to think generationally. The question becomes, not how will the current generation respond to an idea, but how will the idea evolve and adapt with every generation? What values and cycles will be perpetuated generationally?

Now, when I'm asked if I think the commercial space industry will succeed, I often remind the person asking of cycles of prioritization and deprioritization we have seen in the past. Of the Concorde and Aramis and of the gradual defunding of the Apollo program, which stemmed from rising public unpopularity due to astronomical costs and continued civil unrest.* I think of how my students—a healthy mix of STEM majors and social scientists—shake their heads no when I ask them if they think we should be going to space, and remember that *they* are the next generation of policymakers and world leaders. Then I wonder how sustainable an industry of start-ups with a desire to colonize, extract, and develop—rather than conserve and maintain—really is.

*Then* I think about what innovation scholars have been arguing as of late. That a lot of what is being referred to as "innovation" in the twenty-first century is actually the same technology being redesigned for the same technology sectors. That innovation has plateaued since what was essentially the Apollo era and that technocrats have been

---

* One might also argue that outer space exploration never completely fell out of favor in the 1970s, but it's worth noting that funding dramatically decreased.

reinventing the wheel for a very long time.[9] That when we think about innovation in relation to industrial-era innovation, which was spread out across all sectors of human society, innovation today has contained itself to incremental shifts limited to the communication sector.

Some of the same innovation scholars have also commented on prophecy in relation to technological innovation and futurism.[10] One of my favorite innovation scholars (indeed, such a favorite that he's on my dissertation committee), Lee Vinsel, went as far as to say that "even simple historical reflection shows that humans are truly terrible at predicting the future, especially technological futures."[11] He went on to argue that a prediction is only "useful if it encourages us to deal with problems we've been ignoring for generations, and harmful if it encourages us to believe that these problems will solve themselves, or that technology will solve them for us."

Let's also hold on to that idea as we move forward. What problems does human space exploration solve? And the prophecies of commercial space companies . . . their visions for the future . . . do they encourage solutions for issues on Earth?

*What values and beliefs are being prioritized as we consider human space exploration?*

------------

I woke up the morning after my drunken rooftop venture with a pounding headache, an intense craving for phở, and a remarkably revived motivation to get my shit together. I had wanted to be an anthropologist since third grade. Now was not the time to grow fearful of my own shadow. I needed to sort out *my own* priorities and commit to them.

Two weeks later all of my applications paid off. I had an internship offer with the Commercial Spaceflight Federation.

# 8

# CELESTIAL MOTIVATIONS

*The Robert H. Goddard Memorial Dinner, Washington, DC, 2019* CE

**To say that Washington, DC,** is a hub for the commercial space industry would be a massive understatement. There is a thriving space community centered in the nation's capital, which attracts organizations from across the globe. I've been told by friends based out of other space industry hubs (Cape Canaveral, the Bay Area, Houston, and ESA bases, for instance) that you can't find anything like it elsewhere. I think a major source of the uniqueness of the DC space community stems from a sense of camaraderie that it encourages. This camaraderie blossoms through extensive extracurricular networking. For example, the DC space industry hosts a plethora of social events, such as Space March Madness, SpaceBall at Nationals Park, Space Oktoberfest, and Space St. Patrick's Day. And there's my personal favorite, a recent addition in honor of pride month, Gay Space Happy Hour,* which was held at everyone's favorite local LGTBQ+ friendly bar. The city also hosts the Robert H. Goddard Memorial Dinner each year (known colloquially as Space Prom) and its corresponding gala *des refusés* for

---

* I have no idea if this was the formal name for the event, but this was what members of the space industry colloquially referred to it as.

those who don't want to pay for the thousand-dollar ticket to attend the official event. Space industry employees also attend museum events together, and political events. They travel to a plethora of conferences each year and participate in their corresponding after-hours itineraries.

"Everyone knows you go to Space Symposium for the after-parties, not the panels. That's where the real deals are done," I was told once by a colleague.

What's more, members of the more than eighty space-related companies that frequent the city just *hang out* with each other. They form cliques and social groups. Throw extravagant birthday parties and invite the whole industry. They go ice skating in the winter and bowling in the spring. They grab drinks and gossip and rock climb and host baby showers and bachelorette parties and barbeques. Their camaraderie goes far beyond workplace exchanges.

The DC space industry is a hyper-connected microculture, deeply involved in government relations, with access to *a whole lot of power.*

And when I say *power*—just to clarify—I mean I ran into the Secret Service more than once on the job.

The shift from working as an anthropologist in the New Mexican desert, at the fringes of space society, to working as a space policy intern at the heart of the most elite part of the industry was as overwhelming as it was transformative.

With nothing more than two suitcases and a minimum wage internship, I booked a long-term Airbnb in the southeast part of the city, just below the Anacostia River. The area has a notoriously high crime rate and is known for issues with gentrification.* I remember one advertisement for the housing listing I chose read something along the lines of "Really lovely place. Wonderful host. Did get mugged on the way home once but it was not the Airbnb's fault. Otherwise, ten out of ten." Unbothered, I booked the place, packed

---

* In Anacostia, crime, including violent crime, is reported as 38 percent higher than in the rest of the District of Columbia.

my heels in my bag each day, and wore tennis shoes to walk home from my strangely upper-class job, in case—gods forbid it—I needed to revisit my high school fence-hopping ways as a protective measure at some point. Though there was at one point a shootout on my doorstep that scared a poor Uber Eats driver half to death, my time in Anacostia remained otherwise uneventful.

It wasn't long before I discovered exactly what space industry work I had signed up for. The Commercial Spaceflight Federation, the organization that I would write space policy for, performed business development, government relations work, and policy research for over eighty commercial space companies, including the most well known (Blue Origin, SpaceX, Virgin Galactic, etc.). They worked directly with the executives of these companies. The presidents, directors, founders, board members and other leaders in the industry. They also worked directly with astronauts, celebrities known for their involvement with space activities, government representatives, and government space organizations (NASA, the ESA, etc.). To use another cheesy space metaphor, I had just shot from low Earth orbit into uncharted starry waters, and to top it off, I knew absolutely nothing about business development, government relations, or policy. Many organizational anthropologists do specialize in these areas, and as I worked to complete my doctorate, I would gain expertise in policy and government relations, but at the time, my training was as a social scientist and historian.

Honestly, I think they hired me because Jane Kinney, their director of business operations (who, several years earlier, connected me with Spaceport America), had interest in my skills as a writer and as a cultural expert who researched space. As much as she and I would butt heads at times, I was grateful for her support, because it was the first time I felt really valued for those skills.

But not everyone understood or appreciated my background. Though I did make many good friends during my time in the nation's capital, reactions to my field of study were incredibly mixed, and many of the individuals I spoke to either didn't fully comprehend what an anthropologist's role was in the space industry or didn't take it seriously. Many equated outer space anthropology to researching aliens or thought social research on outer space was valuable but not valuable enough to hire. Or, if they did want to hire me, they weren't sure where I would fit. Do you put an anthropologist in marketing? Business development? Government relations? Policy?

"An anthropologist?"

"Yes. Who studies outer space." During this particular encounter I was at the Robert H. Goddard Memorial Dinner. Against common odds, I had acquired an *actual* ticket via a friend in the industry and wasn't attending the tandem party for those without admission.

"But how? On archaeology sites?" The individual asking the question would go on to be a prominent leader in the Space Force after it was established in late 2019. Together we floated among ornate table displays, primly dressed wait staff, and a swarming huddle of partygoers clustered around Buzz Aldrin. My colleague listened as I described exactly what a "space anthropologist" did.

"Well," I began slowly. "Not all anthropologists dig up bones. Social anthropologists can research anything really . . ." I waved a hand at the room around us. "Grand parties. The people serving them. Sommeliers. Engineers. Astronauts . . . The military. You name it."

His eyes widened in surprise before he began to laugh. Like many of the members of the military attending the Goddard Memorial Dinner that evening, he was wearing full dress uniform, and the awards pinned to his chest bounced with each exhale.

"But what do you do? Observe these people?"

"More or less," I nodded, shifting in my heels. Three hours into the event and I had an awful blister.

"But then how do you research outer space? You can't exactly go."

I frowned. He sounded rather sure of himself.

"Anthropologists research space by analyzing human interactions with the cosmos here on Earth. We contextualize the values and beliefs placed on human space exploration. Study how knowledge is created, and how expertise is labeled as expertise. For example, Janet Vertesi worked for the Jet Propulsion Laboratory and studied how the rovers *Spirit* and *Opportunity* were anthropomorphized and embodied by scientists, and Lisa Messeri researched how Martian maps created by NASA framed Mars as 'democratic.' We take those values and beliefs and translate the ways in which they develop into policy and . . ." I took a breath. "We explore *up* by looking *down*."

He grinned, bobbing his head excitedly. "That's wonderful! But how do you turn that into *actual* research?"

I tried not to choke on the sip of Merlot I just had, counted to ten, and imagined jettisoning this man out an airlock. He meant numbers. Like many in this industry, he was not quite sure how one could turn qualitative, social data into practical results.

I swallowed slowly.

Very slowly.

"If you don't mind, I think I'm going to try and get Buzz's autograph for my dad while we're near him. Why don't I give you my card and send you some links to explain further?" My offer was genuine. It was only in the moment, overwhelmed by the social minefield that is a gala, that my patience wore thin. Days later, the two of us would correspond over email, exchanging questions and answers about each other's fields of work.

But there at the gala, we traded embossed slips of paper in the traditional way. His showcased a gilded version of his military affiliation, while mine featured the profile of a robot whose story I will tell later. Then we separated, drifting in opposite directions.

I didn't risk the crowd around Buzz Aldrin, knowing I would never make it to the front before the gala was over. It's true that my father would have loved Aldrin's autograph, having been one of many children who sat with bated breath in front of a black-and-white television set to watch man's first steps on the moon. But honestly, rather than get the famous astronaut's autograph I had the urge to save him from the swarm. No wonder he didn't stay at these events past eight. I could only imagine how overwhelming it was—the mob of fans around him was huge! I sighed and headed toward the Planetary Society's table, looking forward to some friendly faces. On the way, I moved past my employer's table, swimming through a sea of ball gowns and tuxedos, barely keeping my head above all the chiffon. Later that evening I would Uber back across the river to my Airbnb in Anacostia and be reminded yet again, as I hung my own gown in my tiny rented closet-sized room, of the massive wealth disparities I was not just bearing witness to but playing a part in.

The DC space industry was indeed a social minefield, and the hyper-professionalism that shadowed the nation's capital didn't suit me. Business casual felt not so dissimilar from a high school uniform, and I remain strongly of the opinion that intelligence is not contingent on appearance. But it certainly felt like my blue hair and tattoos were commented on far more than the content of my work. Cultural training had long since taught me that the social etiquette of the nine-to-five world has been founded on traditions of sexism and ableism, and it didn't help that at the time the Commercial Spaceflight Federation had issues with micromanagement and overworking their interns (now hopefully resolved with completely new staff). In the shade of such things, my creative spirit and inquisitive nature curled in on themselves. Not unlike how they had in high school. I felt conflicted constantly. Hadn't I wanted to work in the space industry? Isn't this what I had been working toward? Didn't I want to find a way to marry the worlds of anthropology and space science?

The longer I worked in the commercial space industry, the more comfortable I became with my nonconformity. Sure, complete rebellion against space exploration still scared me half to death, but perhaps "industry work" just wasn't for me. Perhaps there was a place—elsewhere in the big wide world—where I could research space from a more ethical and cultural standpoint and feel valued for doing so. Perhaps that place was academia. I wasn't sure. But the longer I spent writing space policy, the more comfortable I became with the thought of decolonizing human space exploration. I just didn't know that was what I was beginning to do.

Now, some of you might balk at denial of the nine-to-five tradition and say, "But what alternative to these structures are there?" Actually, I am not the first researcher to argue against the traditional workday and socially constructed standards of professionalism. Consider the fact that just two decades ago it was unacceptable for women to have bare legs in the workplace, and it is still argued by many workplaces that humans cannot be productive in sweatpants or after the hours of 6:00 PM.

One day, I hope they will call such standards as silly as the mandate of '90s panty hose. After all, it took one pandemic to teach us that the majority of white-collar jobs can be performed from home. Hopefully it does not take another to strike down stereotypes about tattoos, hair color, and sense of style, not to mention other phenotypic attributes that have been causing workplace lawsuits and much more egregious civil rights issues for far too long.

But I digress.

Social minefield aside, my job at the Commercial Spaceflight Federation gave me direct access to observing and participating in the construction of space industry beliefs and values. In a political context, I watched my colleagues and friends lobby for commercial space interests. I attended any and every congressional hearing related to outer space. The Commercial Spaceflight Federation hosted and attended events with congressional representatives and senators interested in outer space. We wrote white papers and press releases on relevant space policies. We produced policy regarding commercial space standards, regulation, export control, safety, small satellites, and spaceports. In an economic context, we reviewed NASA's annual fiscal budgets, kept on close terms with space economists, followed commercial space stock markets, and tracked government funding.[1]

In other words: I observed, firsthand, the process of political and economic *prioritization* in the space industry.

This prioritization begins with the establishment of mission statements: sets of principles and goals created as the bedrock of public and private ventures—created to realize specific intentions, beliefs, and values. These intentions, beliefs, and values stem from cultural and historic traditions, and through organization develop into constructions of knowledge and expertise, which then result in technological and scientific development.

Or, in simpler terms, scientific processes and technological artifacts are socially constructed. They are products of knowledge systems designed based on human interests and values. Scholars in science and technology studies have created a whole field to understand this better.[2] Looking back, we can think of the Concorde, or Aramis, or the creation of Spaceport America as examples of this. These were technological artifacts or zones created to fulfill certain political interests, values, or beliefs. Examples of these values, beliefs, and interests include a desire to expand, to conquer, to progress, to surpass, and to transcend preexisting modes of transportation and human organization. Mission statements surrounding these artifacts and zones lay out these efforts prior to their enaction.

Mission statements are modern-day prophecies, and if we explore the mission statements of prominent commercial space organizations, we can start to see not just what they prioritize but how these statements extend harmful colonial and imperialist legacies.

SpaceX's mission statement speaks to making humanity multiplanetary, developing reusable launch vehicles, reducing the cost of human space exploration, maintaining a safe and sustainable orbital environment, returning astronauts to the moon, and promoting commercial crewed programs.[3] Axiom Space's mission statement advocates for a "thriving home in space" geared toward innovators, governments, and individuals interested in spending time in low Earth orbit.[4] Axiom is in the process of developing a commercial space station that they believe would be "universally" accessible to individuals on Earth and would act as the next stage in human progress. The company aims to promote the growth of high-tech industries, stimulate the economy, inspire youth to pursue STEM education, and create stations that will be "a source of national pride and international respect."[5]

Nationalism is a historic and present motivation for the development of space technologies. It plays a large role in commercial space activity. Nationalism is detrimental when it comes at the cost of other nations or peoples. A modern example of nationalistic interests in space exploration would be something like this statement made by the Canadian Space Agency regarding the crew of the Artemis II mission: "We are going back to the Moon and Canada is at the center of this exciting journey," said Honorable François-Philippe Champagne, the minister responsible for the Canadian Space Agency. "Thanks to our longstanding collaboration with NASA, a Canadian astronaut will fly on this historic mission. On behalf of all Canadians, I want to congratulate Jeremy for being at the forefront of one of the most ambitious human endeavors ever undertaken. Canada's participation in the Artemis program is not only a defining chapter of our history in space, but also a testament to the friendship and close partnership between our two nations."[6] And a more historic example would be John F. Kennedy's famous speech given at Rice University in 1962:

> We set sail on this new sea because there is new knowledge to be gained, and new rights to be won, and they must be won and used for the progress of all people. For space science, like nuclear science and all technology, has no conscience of its own. Whether it will become a force for good or ill depends on man, and only if the United States occupies a position of preeminence can we help

decide whether this new ocean will be a sea of peace or a new terrifying theater of war. I do not say the we should or will go unprotected against the hostile misuse of space any more than we go unprotected against the hostile use of land or sea, but I do say that space can be explored and mastered without feeding the fires of war, without repeating the mistakes that man has made in extending his writ around this globe of ours.[7]

Neither statement seems outwardly harmful, but Kennedy's speech was overcast by Cold War tensions, and like most Apollo-era space exploration, was framed around narratives of prestige and imperialism. His speech directly uses terms like *rights to be won* and stresses the importance of the United States maintaining a position of preeminence in space. If the United States didn't maintain the upper hand in the space race, Kennedy feared an all-out war with the Russians.

The statement from the Canadian Space Agency appears even more benign but still reflects a long history of global international competition. Who gets to be the face of space exploration and who *should* be the face of space exploration? Such questions were at the center of space policy during my time working in Washington, DC, as concerns regarding Chinese and Russian participation in space exploration increased, something I'll return to shortly.

Militaristic themes are perpetuated today across public and private space sectors. The space sector of Lockheed Martin follows three overarching goals to "connect," "protect," and "explore." The company's mission statement focuses on driving innovation, militarization of space-based technologies, "advancing the planet" through space exploration, and developing technologies that will help their member companies accomplish their extraterrestrial goals.[8]

Donald Trump's speech at Cape Canaveral in May 2020 also echoed very explicit imperialist and militarist themes of American dominance, progress through power, and frontier histories. Trump said:

> Past leaders put the United States at the mercy of foreign nations to send our astronauts into orbit. Not anymore. Today, we once again proudly launch American astronauts on American rockets, the best in the world, from right here on American soil. . . . Now these brave and selfless astronauts will continue their mission to advance the cause of human knowledge as they proceed to the International Space Station before returning to Earth. . . . Today's launch makes clear the commercial space industry is the future. The modern world was built by risk-takers and renegades, fierce competitors, skilled craftsmen, captains of industry who pursued opportunities no one else saw and envisioned what no one else could ever think of seeing. The United States will harness the unrivaled creativity and speed of our private sector to stride ever further into the unknown. . . . We have created the envy of the world and will soon be landing on Mars, and will soon have the greatest weapons ever imagined in history. I've already seen designs. And even I can't believe it. . . . The United States has regained our place of prestige as the world leader. As has often been stated, you can't be number one on Earth if you are number two in space. And we are not going to be number two anywhere. . . . Nowhere is this more true than with our military, which we have completely rebuilt. Under my administration, we have invested two and a half trillion dollars in new planes, ships, submarines, tanks, missiles, rockets—anything you can think of. And last year, I signed the law creating the sixth branch of that already very famous United States Armed Forces: the Space Force. . . . To be certain, we

will meet the adversity and hardship along the way. There may even be tragedy, because that is the danger of space. There's nothing we can do about that. The power that we're talking about is unrivaled. There's nothing we can do about problems. But we'll have very few of them. We will confront all of those challenges, knowing that the quest for understanding is the oldest and deepest hope in our souls. The innate human desire to explore and innovate is what propels the engines of progress and the march of civilization. We will preserve and persevere, and we will ensure a future of American dominance in space.[9]

Trump's statement is a blatant example of the more harmful nature of nationalism. His address projected not only American dominance but also a militarized intent.

Sometimes executive administrative goals and priorities are explicitly stated, such as the Biden administration's "Priorities" page on the official White House website.[10] But sometimes, government mission statements slip out in press releases, news briefings, speeches, and of course . . . tweets.

During the Obama administration, slightly different intentions were stated when it was argued that "space exploration is not a luxury, it's not an afterthought in America's quest for a brighter future—it is an essential part of that quest."[11] The Obama administration's primary reasons for prioritizing human space exploration was increased employment, inspiration and innovation on Earth, and human progress. But the Obama administration specifically emphasized that activities in space needed to have tangible benefits on Earth.[12] In his speech at Kennedy Space Center in 2010, Obama said:

> We will increase Earth-based observation to improve our understanding of our climate and our world—science that will garner tangible benefits, helping us to protect our

environment for future generations. And we will extend the life of the International Space Station likely by more than five years, while actually using it for its intended purpose: conducting advanced research that can help improve the daily lives of people here on Earth, as well as testing and improving upon our capabilities in space.[13]

These mission statements and speeches about outer space vary in tone, nature, and intent. But they share an intense passion for human space exploration and consider it symbolic of human progress. They also set the principles and goals of public and private space ventures. Mission statements outline specific intentions, beliefs, and values. These intentions, beliefs, and values stem from cultural and historic traditions, and through organization, develop into constructions of knowledge and expertise, which result in technological and scientific development.

But if we return for a moment to the questions I posed in the previous chapter—what problems does human space exploration solve? And what solutions does it encourage?

Sending humans to outer space is often framed as the solution to issues on Earth. Though I hesitate to argue against the value of the ancillary benefits of microgravity research or outputs such as satellite connectivity, I still question whether the research that needs to be done to benefit Earth has to be done by humans (as opposed to robots). Likewise, I question the value of exploration for the sake of exploration at a time when resources could be, as both the Biden administration and Obama administration have pointed out, directed toward issues of more pressing importance such as climate change, scarcity, and sustainability.

These mission statements might seem harmless at face value. And they might seem like proclamations subject to change with each new political regime and the whims of their CEOs and founders. After all, they don't all have concrete results.

But they are not so harmless.

The statements these public and private entities make can have an impact on technological development. And the past, present, and future of human space exploration as envisioned by space organizations is emblematic of a technocracy—civilization and society, extended past Earthly boundaries, designed around techno-scientific beliefs and values. A celestial landscape where expertise and science reigns and political and economic decisions are based on techno-scientific advancement. This alone is not a dangerous thing, but these beliefs and values, when enacted in the ways that they have been for a very long time, perpetuate not just nationalism and militarism but also colonialism—both in its most explicit form (the space colony) and disguised as scientific achievement.

# 9

# THIS LAND IS OUR . . .

*90.0000° S, 45.0000° E, Antarctica, 1911 CE*

**R**esearch stations have long been critiqued as centers of scientific colonialism, but they've also been revered throughout history as locus points of techno-scientific advancement.[1] In this way, they are a good example of the end result of a techno-scientific mission statement in action.

A historic example of this is the establishment of tropical field stations across the globe from the late nineteenth century through the twentieth. Scholars have argued how scientific outposts created during this time period led toward international scientific recognition and a pathway to understanding the "origins of life" through biological research.[2] But while tropical research stations were used to study diseases and the life cycles of flora and fauna, scientists at the turn of the century also established these research stations in an effort to "effect positive change in the world."[3] This goal was a "civilizing mission,"—part scientific in motive, part spiritual, and part political, not unlike the goals of organizations like Axiom or Lockheed Martin.[4] In other words, the arguments that scientists made were for scientific progress, but these arguments were bound up in territorial claims and ideas that indigenous populations needed to be "saved."

*Scientific colonialism* is when a developed or colonial nation conducts research in a developing nation and then leaves without any investment in human capacity or infrastructure.[5] Marine biologist and Sri Lankan native Asha de Vos reflects on this through her own work to explain that scientific colonialism "creates a dependency on external expertise and cripples local conservation efforts." She adds, "The work is driven by the outsiders' assumptions, motives and personal needs, leading to an unfavorable power imbalance between those from outside and those on the ground."[6]

Techno-scientific ventures such as research stations are rarely called out as instances of scientific colonialism and often go unnoticed and disregarded due to positivist social understandings of techno-science. In other words, it can be hard to remember that scientific research and technological development isn't done with purely benign intentions. Science isn't neutral! It's often conducted with very political motivations at its foundation.

This has been discussed by scholars such as Warwick Anderson, who has commented on the postcolonial impact of scientific and technological institutions as sites for "fabricating and linking local and global identities, as well as sites for disrupting and challenging the distinctions between global and local."[7] Likewise, Michael Adas has written about Europe's extensive history of "civilizing missions" through scientific, technologic, and other enterprises.[8] The research station is an extension of colonialism, both in places like the tropics and in uninhabited places such as Antarctica or outer space, because it can extend these "civilizing missions," harmful legacies of colonialism, and practices of extractivism into uninhabited zones.

Some have tried to argue that space stations are not colonies at all because they aren't "inhabited." Peder Anker reviews these perspectives in his article "The Ecological Colonization of Space."[9] He addresses leading advocate for space colonization Stewart Brand's defense of the term *space colony*, where Brand stated that space colonization was unproblematic since "no Space natives [were] being colonized."[10] In his response, Anker argues, "When space colonies became the model for Spaceship Earth, all human beings became 'space natives' colonized by ecological reasoning: social, political, moral, and historical space were invaded by ecological science aimed at reordering ill-treated human environments according to the managerial ideals of the astronaut's life in the space colony."[11] Similarly, Christy Collis and Quentin Stevens address resistance by some scholars to using colonial terminology in relation to Antarctic colonies.[12] For example Hans Kohn, author of "Reflections on Colonialism"—which was written after the "Heroic Age" of Antarctic exploration in the first two decades of the twentieth century—argued that "because Antarctica was not originally inhabited by indigenous populations, the terms 'imperial' and 'colonial' cannot apply to the activities or attitudes of its current occupants and claimants."[13] Others have claimed that the "emotional connotations of colonization" are irrelevant in relation to Antarctica and that the "artificial nature" of Antarctic stations results in the term not applying.[14] But the work of Antarctic scholar Jessica O'Reilly contradicts these arguments. Although Antarctica *technically* has no indigenous population, it is *not* uninhabited. Antarctic base scientists, staff, and visitors form the Antarctic community, and they wield tremendous power through scientific expertise.[15] Antarctic scientists, and those based out of other research stations, use scientific power as a vehicle for extending the colonial enterprise beyond uninhabited spaces to greater society.[16]

In other words, *people listen to scientists*. And because of this, scientific power is powerful and hard to criticize.

Like Antarctica, outer space remains unoccupied by any indigenous populations, but framing it as undisturbed or as a "commons" has its own consequences.[17] A *commons* is a region or resource available to everyone. The recent 2023 Commons in Space virtual conference* emphasizes the space industry's continued desire to frame

---

* I participated in the conference on a panel titled "Interrogating the Commons: Coloniality, Harm, and Alternative Models for Space Sustainability."

outer space as such.[18] But doing so, it maintains a capitalist and colonial framework of outer space. Designating outer space as a "commons" frames its resources and environment as freely available for the taking.

Colonization also exists in outer space, not only in the form of speculative claims to planets, asteroids, and moons but also in other more tangible ways. Optimal orbits of satellites and space stations are fought over, something which one reporter described as a potentially lethal parking crisis in space.[19] And extraterrestrial resource extraction has been a prevalent conversation in the industry, as this book examines. Resource extraction also overlaid early narratives surrounding Antarctic exploration, until the prospect of mining in Antarctica was outlawed in the Madrid Protocol.

It is not that we shouldn't trust scientists or fund them as they propose the development of research stations; it's that we should be mindful of the motivations behind what is being funded. As neutral as research stations might seem, they are often developed with both scientific and political intent. Remember, research stations also come with mission statements. The history of this can be traced back to the "Heroic Age" of Antarctic exploration. This time period signified an international race to stake territorial claims on Antarctica, less related to scientific values and far more tied to goals of prestige and nationalism. *The Lands of Science: A History of Arctic and Antarctic Exploration*, written in 1921, offers a glimpse into the way Antarctic exploration captured the public imagination and a sense of nationalism:

> It is not only that we meet here with examples of heroism and devotion which must entrance mankind for all time. It is not only that there are dangers to be encountered and difficulties to overcome which call forth the best qualities of our race. These, no doubt are the main reasons for the deep interest which polar exploration has always excited. But there are others of almost equal importance. These

regions offer great scientific problems. They present wide
fields of research in almost all departments of knowledge.
They have in the past yielded vast wealth, and have been
the sources of commercial prosperity to many communities, and they may be so again. Their history is a history
of noble persevering effort.[20]

Markham refers to Antarctic explorers as "true Britons" and
"heroes of the age."[21] The race to the South Pole in the early twentieth century, led by Robert Falcon Scott, Roald Amundsen, and Ernest
Shackleton, carried with it the weight of nationalism and imperialist intentions, masked by a masculine sense of adventure as well as
scientific interest.[22] Eventually—after much conflict—Antarctica was
divided into seven territories, colonized by different nations: Australia (1933), Argentina (1943), Chile (1940), France (1924 and 1938),
New Zealand (1923), Norway (1939), and the United Kingdom (1908,
1917 and 1962).[23]

In 1991 Antarctica became subject to the Madrid Protocol, an
extension of the Antarctic Treaty. This occurred for reasons similar
to why the tropics became a focal point of scientific fieldwork at
the turn of the century. The Madrid Protocol added a category of
environmentalism to the Antarctic Treaty's mission of promoting
"peace and science."[24] Both the tropics and Antarctica were framed
as a crucial resource in desperate need of saving. Another "civilizing
mission." Yet O'Reilly points out that "Antarctic scientific excep-
tionalism—both the tremendous creative potential for obtaining new
knowledge and the exceptions that permit scientists to both participate in and opt out of the environmental policies for the continent—is born from the same environmental imaginaries and historic
legacies of exploration, extremism, and nationalism that frame the
current geopolitical situation for the continent."[25]

To this day, Antarctic explorers and researchers invoke nationalistic claims bound up in historic context and national character to

garner authority on the continent.[26] O'Reilly notes that "the long history of science and colonialism, particularly among the British colonies, continues to play out (and be disputed, critiqued, and rebelled against) in contemporary Antarctic science and policy, particularly in international Antarctic geopolitics."[27] In other words, scientists try to maintain authority over Antarctica as a region and as a space for scientific development. They do so by competing for funding and other forms of scientific credibility.

At one point, the United States and USSR created Antarctic stations in an attempt to gain strategic positioning during the Cold War.[28] During this time, Antarctic scientists and explorers went so far as to claim that Antarctica is the closest analogue to outer space on planet Earth, as a way of convincing Cold War governments to support their research projects.[29]

But also consider not just how access to territories is prioritized but also how rights over a certain environment or location are defined. In the case of tropical field stations, colonial superpowers claimed rights to land that had long been inhabited. But territorial claims-making is part of colonialism, and one of the primary reasons why it is so difficult to unravel narratives of progress from colonial narratives.

During the Apollo era, arguments for space exploration were focused on America as a nation. In Kennedy's 1961 "Special Message to the Congress on Urgent National Needs," he presents the decision to go to the moon as a choice to the nation, and one which will impact the distribution of government resources but also guarantee, he argues, "an affirmative position in outer space."[30] The international scientific stage, mentioned in Megan Raby's book *American Tropics*, which was fought for so ardently through the establishment of tropical research stations and Antarctic stations, had expanded into the solar system.

Right around the time that America put a man on the moon, the tropics and the Antarctic were being framed as precious, biodiverse environments, in desperate need of saving. According to scientists, the world (not just the nation) was facing a serious crisis, and tropical research stations offered a solution.[31] Space exploration would also offer a solution to international crisis. It would help the United States maintain peace in a time of Cold War.

Does this sound familiar?

It should, because it's very similar to the rhetoric on faith, risk, and sacrifice being propagated by commercial space companies that I witnessed firsthand during my time at Spaceport America. And this new rhetoric goes hand in hand with old colonial intentions and positivist scientific perspectives that have not disappeared.

Biologist Danielle N. Lee examines rhetoric that focuses on leaving Earth to "save the human population," whether from ecological catastrophe, overpopulation, or asteroid strike. She argues that these narratives are often exclusionary (who are we saving Earth from and who is being saved?) and that it encourages the public to give up on the Earth.[32] Like I mentioned before, many of the commercial space organizations I worked with focused on outer space as a "plan B" for humankind should Earth "fail" (there's that dreaded word again), and Elon Musk has pointedly discussed this in interviews about SpaceX. But again, I reflect on innovation scholars, and scholars of "maintenance," who speak to the disappearance of cultures of care, cultivation, repair, and conservation, which I see reflected in conversations about outer space.[33] Should we not be caring for the Earth we have, rather than searching for a plan B?

Just as species diversity was reframed by tropical scientists as a solution to the world's economic and environmental problems in the 1960s and 1970s, so too has space exploration been carefully framed as a *solution* to global problems. Words like *innovation* and *progress* are emphasized as motivations and benefits in planetary cost-benefit analysis for why it is worth it to colonize other planets and leave Earth behind. Meanwhile colonial, capitalist, and militaristic motives hide behind the guise of science.

During the Apollo era, the popular narrative was one of supremacy, urgency, competition, warfare, grit, and determination. Outer space was a frontier to be conquered and colonized, much like the tropics and Antarctica had once been, not just in the American context but in the Soviet context.

Soviet space stations such as Almaz, Mir, and Soyuz were deeply politicized offshoots of the Cold War, built with "the intention of gathering intelligence at altitudes where one couldn't easily shoot down the vehicle collecting it," among other reasons.[34] James Andrews and Asif Siddiqi, historians of Soviet-era space exploration, emphasized that the Soviet government devoted enormous resources not only to perform space achievements but also to publicize them in domestic and foreign arenas.[35] Scientific claims were made often as assertions about legitimacy, credibility, and power.[36] Andrews and Siddiqi also address the Bolshevik state's attachment to science and technology, and to the new "tools of capitalism" such as early Russian rocketry, Henry Ford's mass production, Frederick Taylor's scientific management, and the Wright brothers' airplane.[37] They also describe Soviet interest to "remake Russia into a modern state, one that would compare and compete with the leading capitalist nations in forging a new path to the future."[38]

Colonial legacies surrounding space stations continue to perpetuate and mirror violent histories of extractivist colonization policies.[39] Yet still motives stem from the idea of progress through basic science—with emphasis being placed on research in microgravity, which can't be conducted in laboratories on Earth.[40]

With space tourism on the rise, outer space is being framed as both a resource and a frontier. Commercial space stations have now been proposed by five separate commercial space companies: Blue Origin, Axiom, Lockheed Martin, Nanoracks, and Sierra Space.[41] Their intentions for the stations are similar: they aim to increase civilian access to space, support scientific (specifically microgravity) research, and to "ensure humanity's future." Axiom seeks to make space "universally accessible" and frames it as the next stage in human progress, while Blue Origin and Sierra Space's commercial space station, Orbital Reef, is being framed as a "mixed use business park" in space.[42] The company states that "seasoned space agencies, high-tech consortia, sovereign nations without space programs, media and travel companies, funded entrepreneurs and sponsored inventors, and future-minded investors all have a place on Orbital Reef."[43] Though increasing access to space might sound like a good idea, the goals of many of these commercial space organizations perpetuate capitalism-driven value systems. They also have the potential to increase wealth disparities, because, yet again, even though space *can* legally be accessed by all humans, not all humans can afford to access space.

With or without an indigenous population, through the act of establishing a colony, territorial claims are made, and the foundations of a society are built. Scientific colonialism is not just about conducting research without democratic investment in human capacity or infrastructure. It's also about the intentionality and political motivation behind the creation of technological development and scientific research. Or, in other words, prioritization.

Uninhabited spaces are contested spaces, and sources for competition and conflict. The case studies of Antarctica, the tropics, and outer space, are part of an extensive history of territorial arguments, frontier-inspired value systems, national interests, and ideas of prestige, competition, and progress. Research stations are created for a variety of reasons, but to establish a colonial outpost—and stake a claim to territory—remains one of the most prioritized reasons. Reviewing intention and understanding prioritization is only the beginning of the process of assigning value to human space exploration. At some point, belief becomes law. Values become funded. Culture becomes right. We have to examine that, too.

# 10

# THE EXORCISM OF MANIFEST DESTINY

*US House of Representatives, Washington, DC, 2019* CE

**B**elief becomes law.

Values become funded.

Culture becomes right.

Expertise reigns.

And this chapter is not about politics. Let's just start there.

It's about the construction of knowledge. About expertise and how humans formulate understandings of what it means to be an expert. It's about echoes of manifest destiny, imperialism, and colonialism, refracted through congressional and corporate priorities. . . .

In this chapter we explore another example of prioritization, shadowed by the ghosts of nationalist intent, and examine how priorities about human space exploration are legitimized through performances staged via what is commonly known as the congressional hearing. Or, in other words, a meeting of a Senate or House congressional committee, typically open to the public and defined by the government as a mechanism "by which committee members gather information."[1] These hearings are framed like places of exchange but, when performed, fall into place like artfully scripted Shakespearean tragedies. They stake socially constructed claims based on carefully designed values.

Attending these hearings was a large part of my internship. As soon as a relevant hearing was advertised, Tommy Sanford, CSF's executive director at the time, and Eric Stallmer, the company's president, would urge us to mark our calendars.

"This is history in the making," I remember one of them saying when the announcement went out for the hearing on the proposal to establish the Space Force.[2] Something did feel historic about those meetings; there was a grandeur to them. A beauty in the spectacle and performance. Many were held in historic rooms, designed in the classic beaux arts style, located just a short walk from the US Capitol Building. Several of these rooms, located in the Dirksen Senate Office Building or Russell Senate Office Building in central Washington, DC, feature intricate woodwork and marble staircases. Bronze reliefs along the exterior of Dirksen feature man in states of Earthly manual labor. Each relief is emblematic of a different industry: shipping, farming, manufacturing, mining, and lumbering. The stage of these performances is undeniably rich in an austere sort of beauty that does indeed make even the homeliest audience member feel as though they play a profound role in a moment in history.

But let's again be clear that these are performances. The congressional hearing is a place in which issues are raised and questions are answered. It is a forum. A tribunal. In some cases, a live op-ed. But the system of answers, questions, and statements, and even the range of issues chosen to be noticed is based on a very unnatural, culturally biased selection. Who is chosen to stand at the expert witness panel and respond is a careful, intentional process performed by House and Senate committee members and staffers. The Senate Manual explains that committees are entitled "upon request made by a majority of the minority members to the chairman before the completion of such hearing, to call witnesses selected by the minority to testify with respect to the measure or matter during at least one day of hearing thereon."[3] In other words, it's up to the committee to choose the congressional hearing witnesses. The manual itself does not address how expert witnesses should be chosen, but in a congressional guide to Senate hearings, one author said, "In choosing witnesses, committees pay careful attention to the viewpoints that are heard, who should testify, and the order and format for presenting testimony. In some cases, a committee strives to assure that all reasonable points of view are represented, while in other cases witnesses expressing only particular points of view are invited."[4] The federal definition of *expert* remains equally vague when choosing an expert witness. According to Rule 702 of Federal Rules of Evidence, an expert witness

> is qualified as an expert by knowledge, skill, experience, training, or education [and] may testify in the form of an opinion or otherwise if: (a) the expert's scientific, technical, or other specialized knowledge will help the trier of fact to understand the evidence or to determine a fact in issue; (b) the testimony is based on sufficient facts or data; (c) the testimony is the product of reliable principles and methods; and (d) the expert has reliably applied the principles and methods to the facts of the case.[5]

In other words . . . anyone can be considered an expert. Especially when we consider the contested history of other terms that have also been argued to be socially constructed such as *methods*, *evidence*, and *facts*.[6] One might wonder how the questions asked and stories told might change if our definition of *expert* shifted. Consider an Earthly example, before we recenter our orbit back into outer space. Several scholars have studied expertise over the years, and more specifically, *lay* expertise and how the idea of an expert shifts based not just on experience but also on class, race, sex, and other demographic factors. For example, Stephen Epstein studied the role of activists in biomedical research related to the AIDS epidemic in the 1980s and '90s. Epstein looked at how AIDS activists were able to construct their credibility as experts because many of them were white and male. Some were already involved in science or politics. These individuals could speak "the language" of medical science and of legislation and, because of that, be taken seriously in political and technocratic circles. As opposed to, for example, suffragists or civil rights activists who had their sex, race, and, oftentimes, class, working against them.[7] Though both populations were activists, one group's demographics helped lend them credibility as experts.

In another example, science and technology scholar Brian Wynne studied Cumbrian sheep farmers' responses to scientific advice about the restrictions introduced after the Chernobyl nuclear disaster.[8] Wynne's research highlights a push and pull of credibility and specialist knowledge as it is accepted and denied between scientists and sheep farmers trying to understand Chernobyl's radioactive fallout and its impact on local trade. While the scientists had *technical* expertise about the effect of the disaster on sheep and the environment, the farmers had *lay* expertise about the same topic. Both were accurate and valuable, but in some cases, lay expertise revealed intricacies and data that could go unnoticed by technical scientific collection.

Here are some other questions that might prompt new ways of thinking about expertise and who might make a good expert witness in Congress: Who has more expertise on movement in games, a video game designer or a player that's put in two thousand hours? Which expert knows more about how consumers might react to a change in alcohol sale laws, the executive of an alcohol company, a scientist who studies vodka, or a bartender who's worked in bars for twenty years? What is Bill Nye an expert of—science or science communication? (No shade to Bill Nye, by the way. He's a very nice guy.)

How might the identity of the expert shape the way representatives and senators ask questions of expert witnesses? How might demographic bias throughout the hearing process shape not just the way expertise is constructed, but also the way in which its construction is related to class and identity? Phrased differently, how might questions differ when being directed at a sheep farmer versus a scientist?

Perhaps it would not at all.

But inquiries like these also make me wonder how much freedom expert witnesses have in the "performance" of their testimony and the way they respond to the questions asked of them.

I remember when a colleague of mine was asked to testify before the Senate and how panicked they were. How they double-checked every word of their witness statement and had multiple friends in the industry reread it dozens of times to check for "political correctness," tone, and accuracy. Even as an audience member attending congressional hearings, I participate in a performance, dressing and even taking notes in a certain way that speaks to specific definitions of *professionalism* and *decorum*. For example, there are no rules that I know of against having laptops out at congressional hearings, yet I rarely saw one, and thus chose instead to take notes with pen and paper.

The more congressional hearings I attended, the more I started to wonder why the witnesses chosen for hearings on topics related to outer space were being chosen, and why they were considered experts. I also started to think about how the questions being asked of these witnesses reflected the United States' beliefs and values when it came to human space exploration.

"Do you believe establishing a Space Force will contribute to the development of a space warfighting ethos and culture that does not exist already today?"[9]

"Can you assure the US taxpayer that we aren't simply outsourcing space exploration when we have companies designing lunar landers right here in America?"[10]

"So would you say that, as we look to space, is there any special reason not to believe that all of the factors that affect air, land, and sea around our hemisphere, that any of those will be significantly different? In other words, can we not expect at least similar activities as we have similar bad actors or the same bad actors who are already in space, such as China and Russia?"[11]

Determined to understand the root of such nationalist and xenophobic debates, I decided to take a sample of twenty congressional hearings related to outer space between the years 2017 and 2021.[12] The first thing I did was list the witnesses for all of these hearings and the companies that they worked for. Immediately, I noticed several important things. First, there were *a lot* of familiar names. I knew many of these people. I had met with them. Had business meetings with them. Coffee. Drinks. Seen them at those DC space events. And that matters *significantly,* because it was one of the first indicators of how small the circle of individuals was that Congress was pulling from. As well, many of the individuals on the list had spoken before Congress multiple times. Thirty-six out of sixty-five expert witnesses (55 percent) worked for the government in some capacity, whether that was a branch of NASA, the military, or a state-level institution. Out of those thirty-six, nine worked specifically for NASA. Only ten out of sixty-five (15 percent) worked for private organizations, and fourteen out of sixty-five (22 percent) worked in academia.

Eighteen out of sixty-five were women (28 percent).

Almost all of the individuals on the list were in elite job positions. In other words, they were the founders, directors, CEOs, administrators, generals. Not a single expert witness could be considered a lay expert or representative of the public opinion. The only early-career job title I saw was "research fellow" (which, let's be clear, you typically need to be a postgraduate to achieve. So one could consider the position somewhat elite, even if research fellows aren't paid like they are elites).

It was very clear whose voices were being chosen to be heard. And one might say, "Well, the point of congressional hearings is not to hear from the public!" But should it be? Should Congress be hearing from a more diverse set of voices even if the issue is techno-scientific? Is the government undervaluing lay expertise? Certainly, they should be diversifying panels in regard to demographics, but perhaps lay expertise should be taken into consideration when approaching the creation of expert witnesses. Perhaps motivations should also be examined for *why* a person is being chosen to act as an expert witness. For example, lots of astronauts have been on expert witness panels for congressional hearings. But what makes an astronaut an expert at space exploration? Is it because they *went* to outer space? Trained for space exploration? What if they were the pilot on a crewed mission rather than, say, the crew scientist or medical officer? What makes an astronaut the right person to answer questions about, say, the next fifty or a hundred years of space exploration?[13] Are commercial astronauts equally qualified to answer these questions?

Don't get me wrong. I am not denying that astronauts are experts; they are—in the incredibly niche things that they are trained to do. But why does Congress think that astronauts are the best voices for answering questions about the future of humankind in space? My point is to deconstruct these motivations because the minute we break down why values have been placed on expert witnesses in the past, we can start to see how value might be placed on an alternative set of voices, and why these voices should be included in conversations about human space exploration. These voices could be lay experts, citizen representatives, or voices from industry that have yet to speak before congress.

———————

Chin in hand, I tilted forward to listen closely to the questions representatives asked expert witnesses. Some of these did align with my own research and musings:

"So what would you say to people who submit that we should, rather than spend money on human exploration, or astrophysics, or planetary science, spend it in other areas? What's the best response to that?"[14]

"As former NASA chief technologist, can you discuss some other notable scientific and technology advancements that NASA has played a part in and how they have helped to shape the lives of everyday Americans?"[15]

"How can NASA and the community both encourage ambitious breakthrough science while minimizing unanticipated costs and delays that may come with pushing the edges of innovation? And must pushing the edges of innovation and discovery always be equated to large and expensive missions?"[16]

"Just what assurances can you provide that we're not reinventing the wheel but we're adding value?" (on the Space Development Agency).[17]

"What are the primary goals and objectives for going to the moon? Are they geopolitical, scientific, commercial, or as risk-reduction efforts for an eventual Mars mission? On which goal is NASA basing its architecture and mission decisions?"[18]

"So, $500 million a year, half a billion dollars a year, in organizational change. I mean, are you coming before us saying, 'We can't manage this now, and we need to spend half a billion dollars a year'? You understand what I'm asking, I'm sure. Convince me that this makes sense.

That it's worth $500 million a year" (on the creation of the Space Force).[19]

One thing that is interesting in hindsight is that outer space has such a bipartisan appeal that pro-capitalist and pro-caretaker space philosophies are not split directly down party lines. Interest in space exploration does tend to have a regional focus—for example, representatives of states or cities with high investment in space tend to promote space exploration more—but space capitalism is not specifically Democratic or Republican.

At congressional hearings, questions by representatives also made clear the imperialist and militaristic motivations and priorities that still shadowed human space exploration. Intense political focus was placed on national security and military dominance.

"Are we in a position where we are now acknowledging that, as a domain, we have to have the same types of capabilities, both offensive and defensive capabilities, or are we restricting ourselves right now to defensive capabilities only?"[20]

"Excuse me, Mr. Whiting, in your opinion, has space already been weaponized by countries like China and Japan? And what do you make of the satellites that reportedly shadow other satellites?"[21]

"Have the Russians been told that this behavior is unacceptable? And if so, what was their response?"[22]

"So, again, can you assure Florida's taxpayers that NASA will not outsource its space exploration and will continue to focus on American companies, as it has throughout its history?"[23]

"There's no way to avoid space being central to our way of war, is there? I mean, some of it is a legacy based on our technological advantages, going back to the early days of the space era, but it's also just the fact that we live in the new world, and they all live in Eurasia, and we have to project power across a global scale, which depends on space."[24]

"Chinese satellite, eight thousand pounds. They took target practice on it. A week later they had to move the ISS again. Why'd they have to move it this time? More space debris. Chinese again? No, Russian space debris. Well, why would the Chinese and Russians shoot their own dog? Just to prove to themselves, and the rest of the world, that they were capable of doing that. If they can take their own satellites out, they can take our satellites out. So the, you know, the question that begged for an answer is, you know, what are we doing about it, and how can we make sure that it gets addressed?"[25]

I left many of those hearings on edge and feeling chilled to the bone. Spring in Washington, DC, was brutal, but it was not the wind whipping down the city's wide streets that made my muscles tense and had me pulling my coat tighter. Conversations about outer space that sought to *take*, extract, and expand with no thought of the repercussions that might stem from such actions had me apprehensive and restless about human space exploration. What's more, I remained confused as to why so many in the space industry—and the nation— seemed immune to the intense Eurocentrism and xenophobia that layered conversations about space exploration. Though some of my concerns were confirmed by representatives in Congress, I couldn't help but continue to feel isolated by my doubts.

---

"Tommy? Can I ask you something?"

It was the week of the annual FAA Commercial Space Transportation Conference, cohosted by the Commercial Spaceflight Federation, and I had swung by CSF's office late after we closed to pick something up. Tommy Sanford, the company's executive director at the time, was still in his office, working late, when I entered. Typically, I did my best not to bug him while he was working. Our office was modest in size, and it could be easy to distract each other. But this time it was he who came out, stretched, and struck up conversation with me while I collected things for the conference.

Tommy broached the subject about the social sciences and outer space first and spoke with ease that night, about history and the ethics of space exploration. I, on the other hand, toed around the conversation. Like I mentioned before, I was uneasy about bringing up decolonial ideas and the ethical side effects of human space exploration in front of my employers. I worried I would ask the wrong question or interrogate an ideal held too close to the heart. My internship would not last forever. I needed a job, and my interactions with space capitalists in the commercial space industry continued to make me feel like I was in the wrong for questioning celestial motivations. People loved space! Human space exploration was clearly the next step in our species' evolution! . . . wasn't it? . . . wasn't it? What alternative could there be? I certainly hadn't been shown any yet.

But something about talking to Tommy that night revived the curiosity within me that I had been doing my best to stifle. And the confidence in his tone, the way it felt like for the first time he was speaking to me like an equal, the way he acknowledged the social and cultural elements of space exploration, made me wonder . . . What if he understood? What if he thinks about all these consequences and socioeconomic ramifications, too?

What if I'm not alone?

Tentatively, I extended a botanical tendril—a questioning vine—and described the Elysium Effect. The idea that asteroid mining and other space activities might have negative ramifications for people on Earth. That these activities might extend cycles of poverty into space. That wealth disparities and elitism decided whose voices were being included in the conversation and whose were not . . .

His tone changed immediately. Walls went up. A defensive line drawn deep in the soil. And like *Mimosa pudica*—the shameplant—I curled in on myself again, wondering if I was in the wrong. But I tried, before the conversation ended, to understand one final thing.

"Tommy? Can I ask you something?"

He nodded; we were still standing in the middle of the empty office.

"Why do space companies . . . the government . . . hate China and Russia so much? Why does it matter if they get to space first? Would it really be so bad?"

I wasn't naive enough to be blind to the consequences of international relations gone wrong. I knew enough about the Cold War and Apollo-era conflict and many other moments of military conquest in human history to understand the gravity of war. Remember, one of my childhood obsessions was military history, and I would specialize in Cold War history throughout my doctoral studies. It just felt like the *tone* of conversations about Eastern European and Asian countries' role in human space exploration was antagonistic to say the least.

The response Tommy gave me could have been a quote from one of those representatives whose questions made me cringe. Xenophobia with a weighty dash of colonial intent. The answer, in a very traditionally American way, seemed to be to fight violence with violence.

Tommy, to come to his defense, was simply repeating what many others in the industry were saying. These were worries founded on nationalism and militarism. Rather than an acknowledgment of politically systemic issues and how they might be resolved through peaceful methods, the American space industry wanted to maintain celestial prowess first and foremost. In other words, the thought of constructing new knowledge systems, governmental structures, and a future that did not aim for the "development of a warfighting ethos" in outer space had not crossed anyone's mind.

# 11

# THEY MAY NOT BE MAN

*National Portrait Gallery, London, 2017* CE

**C**ongress was not shy about its priorities when it came to human space exploration. Displayed clearly in witness statements and opening speeches at congressional hearings were echoes of Apollo-era nationalism, colonialism, and militarism, resurrected and refreshed. America would be going to the moon again. We would remain the first and only country to do so. Then we would go to Mars. A human presence in outer space would be maintained at all costs. These were facts, not up for debate. The intricacies of the how and when and why were what was being questioned. Beliefs were becoming law. Culture was becoming a right to be enforced, and values were being funded.

But that last part was the bit I wrestled with the most as I tried my hardest to empathize with why space exploration, specifically human space exploration, was so valuable. If an argument was not being made about space exploration for the sake of exploration, or for maintaining the upper hand in stellar warfare; it was often argued that human space exploration was most valuable because of its techno-scientific benefits back on Earth. Entire sectors of space agencies work to ensure that space science has practical, ancillary benefits for our home planet. NASA, for example, has the Technology

Transfer program, which seeks to ensure that "innovations developed for exploration and discovery are broadly available to the public, maximizing the benefit to the Nation," and the NASA Spinoff archives describe some of the commercial products that have stemmed from space exploration.[1] There has also been a tremendous amount of work on the benefits of microgravity research, such as drug discovery, nanotechnology, materials science, tissue engineering, agriculture, medicine, Earth observation, technology development, and more. As well, a long history of technological developments have stemmed from space exploration, such as GPS, weather technologies, minimization, and even memory foam.

The more I understood the benefits of space science, and especially its practical, Earthward observations, the more I rooted for it. But I continued to question how much work in outer space had to be done by human beings—as opposed to robots, rovers, probes, etc.—and I also questioned what research was being prioritized. Of course, outer space exploration is not the only controversial enterprise that the United States funds.

The United States has long been criticized for a defense budget that is larger than the combined budgets of the next nine top-spending countries (one report from 2022 displayed America's $801 billion in defense-related spending next to the combined $777 billion of China, India, the United Kingdom, Russia, France, Germany, Saudi Arabia, Japan, and South Korea together).[2] And the political controversies surrounding defense and space are in many ways related. American beliefs and values related to defense and, more accurately, offense, have influenced things like the funding, approval, and creation of the Space Force. But when examining why things get funded, again we have to ask what problems human space exploration is solving and what solutions it is encouraging for issues on Earth.

Exploring where the money is going within federally funded space agencies isn't difficult. In 2019 NASA released their first-ever agencywide economic impact report, which highlighted the influence of space science on the American economy. The agency's annual budgets are also available to view online at essentially every stage of their processes. What they reveal is that as political beliefs and values have shifted with different administrations, so too has the funding of certain scientific ventures. For example, throughout his administration, Trump cut funding to NASA's Carbon Monitoring System and was accused of restricting communications related to climate change.[3] Meanwhile, he redirected support to the creation of the Space Force, commercialization of low Earth orbit, and a trip to the moon. Likewise, between the Obama administration and the Trump administration there has been a de-emphasis on Mars and increased interest in the moon. All of this is reflected in NASA's yearly budgets (which the Planetary Society does a great job in breaking down for the public each year) and is reflective of political and social interests.[4] Budget data is *not* available for most *commercial* space organizations, because they are privately owned.

But I'll admit, it is strange to witness these shifts from within the industry. I often found myself reminded of George Orwell's *1984*, in which Winston—the novel's main protagonist—experiences how his dystopian world has always been at war.[5] The nation of Oceania is at war with Eurasia and in alliance with Eastasia, and it is never acknowledged that these powers were ever aligned differently. However, Winston recalls that Oceania was previously at war with Eastasia and in alliance with Eurasia, and then suddenly without explanation, the telescreens declared a different enemy. One day they were at war with Eastasia. The next Eurasia. And everyone just pretended it had always been so.

One day we were going to Mars and climate change was a serious issue.

The next day we were going to the moon and climate change was not a serious issue.

And everyone in the space industry gamely went along with such utterings, because that was where the money was going.

I, however, found it very hard to keep up and continued to wonder how trips to either extraterrestrial body benefited humankind on Earth. Still overshadowed by nationalism and a drive to colonize, the motivations behind such ventures felt anything but "scientific." Plus, why send humans at all? Could we not send robots or rovers if such endeavors were truly necessary?

In recent years, astronomers Donald Goldsmith and Martin Rees have compared the costs and benefits of human space exploration to unmanned exploration.[6] Goldsmith and Rees explore whether the potential of machines in outer space is equal to that of humans. To get to the root of the question, they do exactly what I've done throughout this book: first, they examine the motivations for sending humans to space, and then they analyze the pros and cons of unmanned space exploration. They look at the most basic assumptions about space, such as curiosity, exploration, ideas of humanism shaped through discovery, progress, and military positioning, to understand why it is so specifically important to send humans (as opposed to robots) to the stars.[7]

In doing so, they determine that we do *not* need astronauts as space explorers.[8] What can't be done now by humans will be possible in the future as technological advancements continue to be made. And with thoughts of safety, economics, and patience in mind, space exploration should favor automated explorers. Sure, humans might conduct certain geological work faster or with more accuracy, but there is no need to rush these things. Rushing comes from the fear of Earth "failing" and the desire to find a plan B, which as previous chapters describe is a flawed way of approaching Earthly issues in the first place, because it perpetuates narratives that give up on the maintenance of planet Earth. We can't go into space exploration thinking that Earth is a lost cause and that repair is not an option, nor should we prioritize disasters set to occur in distant futures such as the death of our star—yet another argument for interstellar space colonization. Instead, a decentralization of self needs to occur. A dissolution of anthropocentrism, so to speak. And a recognition that future space explorers may not be man, even if man moves through them.

———————

Prior to moving to London, I visited the city as a tourist and went to the National Portrait Gallery at a very significant moment in time. It was the first time that a nonhuman was featured in the gallery: Erica, the robot.

You see, the traditional definition of a *portrait* is a painting, drawing, or photograph featuring a human, and Erica defied these rules. She wasn't human. She was an android designed with a strong emphasis on communication and emotions by Japanese researchers based out of Kyoto University. But I didn't realize that at first. I was simply captivated by her coy gaze, and those shy eyes. The way the photographer, Maija Tammi, had captured movement and mischief and the slightest hint of romance. A secret kept. A backward glance. And then I noticed that Erica's portrait was titled *One of Them Is a Human*, and I read the write-up and learned that she was anything but. Erica was a robot.

Though Erica has never been to outer space, and probably will never go, it feels hard not to reflect on the importance of her inclusion in the National Portrait Gallery when considering the growing importance of robots in human society. Robots already permeate our world in little ways and big ones. From artificial assistants like Siri and Alexa, to curious household helpers like Roombas, to companion robots in hospitals and delivery drones, to rovers on Mars. They have been taught to be psychotherapists and friends. To laugh, respond, and feel shame. To make love and perform certain types of labor. To dance. To diffuse bombs. To, as sociologist Janet Vertesi argued, become extensions of our Earthly selves as they travel across Martian landscapes.[9] Is it a selfish thing to want to send flesh and blood to outer space when humans could continue to perform labor through such beings in a safer, cheaper, and more efficient fashion?

More progress could be made via unmanned space exploration if more resources were allocated in that direction. In other words, a shift in cultural values that acknowledges unmanned space exploration as not only just as valuable as human space exploration but also safer, cheaper, and perhaps *more* valuable. A reprioritization.

Robotic explorers can look farther and see more than human beings ever could on their own. Truly expanding the limits of space exploration is only a matter of setting aside often ego-driven motivations for sending humans out of Earth's orbit. And we can go one step further and acknowledge that perhaps exploration for the sake of exploration is not the destiny of humankind. Prophecies and destinies don't have to span galaxies. They are just as important when focused homeward.

But in the exact same way that techno-scientific installations extend imperialism into uninhabited spaces, automated beings can carry with them colonial contestations. In fact, Vertesi's anthropological work on *Spirit* and *Opportunity* proved just how much of our humanity we leave behind in rovers, and many other anthropologists who study robotics have made similar observations.[10] Leaving space exploration in the hands of robotic colonists should not be taken lightly, because they will indeed be "colonists incarnate," bent to the will of the humans that guide them.

Even a truly artificially intelligent being—a creature able to make independent decisions and improve upon itself through machine learning—would remain affected by the goals and priorities of its designers. What has this being been created to do? Does it explore? Extract? Build? What problems does it solve? What solutions does it encourage?

# 12

# THE DEATH OF OUTER SPACE DREAMS

*Kirwan's on the Wharf, Washington, DC, 2019* CE

**W**ith congressional hearings under my belt, space industry budgets ana-lyzed, and policy written, my time at the Commercial Spaceflight Federation was coming to a rapid end. My internship was scheduled to come to a close in May, and I—like many of the other interns in the DC space industry—was desperately trying to figure out my next step. I think a lot of people assume that this just means filling out job applications and hoping for the best, but it doesn't. Especially not in a line of work as niche as mine. I was hardcore networking. My schedule was absolutely full of coffee meetings and dinners and happy hours. I had a stack of business cards on my desk a hundred deep and just as many LinkedIn connections to follow up on. If anything, the process felt strangely reminiscent of my fieldwork at Spaceport America. And it ended up being strangely similarin other ways, too—ways that I could never have anticipated.

---

The commercial space industry, like many STEM-oriented industries, has an issue with diversity. Males vastly outnumber females, to the point that one female senior colleague once told me, "Next time you're in a meeting . . . count. Count how many women there are. Then count how many people of color. Count how many of these individuals are in executive positions." Once I started counting, the numbers were chilling. Not *once* did women ever outnumber the men in the room, and they rarely surpassed a quarter of the population of a meeting. The numbers were even worse for people of color, nonbinary individuals, or members of the LGBTQ community (if I was aware of their sexual orientation). The lack of diversity is important to acknowledge because when the development of outer space technologies comes only from a specific demographic (primarily the white, male, and wealthy), only one perspective and set of cultural values and beliefs are getting integrated into the design of technologies and implementation of science. Sure, you might argue that these white males come from many different states or countries, but female and nonbinary and LGBTQ and disabled and Indigenous perspectives from these states and countries are also needed.

This is old news, unfortunately, and thankfully there has been a good push across the space industry to increase diversity and inclusion in all ways possible. But what was *new* news in 2019 was that the commercial space industry also had an issue with sexual harassment.

Don't get me wrong. I'm sure these issues started long before 2019, but during my internship, stories started circulating about incidents of harassment, and in the following years, those incidents culminated into an absolute storm of criticism across Twitter (now known as X) and the media.[1] Headlines proclaimed that the commercial space industry's harassment issues were too toxic to fix and that space organizations were ignoring harassment claims made by interns.[2] By 2021 books like Chanda Prescod-Weinstein's *The Disordered Cosmos: A Journey into Dark Matter, Spacetime, and Dreams Deferred* would highlight the widespread impact of sexual assault, gender, and harassment in the study of the sciences.

In an ironic twist, anthropological training does actually prepare the researcher for sexism and issues of gender and sexuality. Different cultures approach these social intricacies in diverse and unique ways. Anthropologists can't assume that the culture they are walking into will share their beliefs and values about gender, sex, feminism, and equality. There are long-standing examples in the history of anthropological research of an anthropologist's gender or age or reproductive status causing complications for their research. For example, in one story, an anthropologist wanted to research infertility in a village but accidentally got pregnant right before she started her fieldwork. Because of this, the community of women she wanted to study refused to accept her, and she had to reorient her research away from the study of fertility.

But, dammit, I was in Washington, DC, and I wasn't even conducting research! I was simply trying to make my way in the world as a social scientist. Yet I too fell prey to moments of harassment—some far more extreme than others.

———————

It's not uncommon for people to ask to touch my hair—especially if it's blue or green or pink. Sometimes I even say yes. Especially when the person asking is under the age of ten and begins her request with "Mommy, look, mermaid hair!" But at the time of the request in mind, my hair was neither an abnormal color, nor was the director at Lockheed Martin I was meeting for dinner under the age of ten. In fact, my hair was a fairly natural brown tone, styled amiably for the meeting, sure, but certainly not worth any exceptional attention.

Also, usually the request to touch my hair isn't followed by "Well then . . . would you like to come back to my hotel room and *just hold me*?"

His was.

I arrived at Kirwan's on the Wharf, where we were meeting for dinner, early, and spent some time walking along the marina alone.

A warm wind flowed off its surface, and I remember feeling good. Confident. If there was a harbinger of the events to come, I certainly didn't see it. This would be my second meeting with this director, and he wasn't the only person at Lockheed I was talking to about job prospects. But he seemed unusually interested in me, and willing in some fashion to figure out the best fit for me within the company. I learned that he had a history of helping interns work their way up into higher positions, and it felt like he wanted to help me maintain my status as a social scientist while working for the company—a rare thing, indeed. It felt, as my internship came to a close, that things were finally falling into place.

I inhaled deeply before entering the Irish-themed pub situated on DC's wharf. The salt in the air was so much denser here. You could taste it in the local oysters and seafood, and it reminded me of my childhood by the sea.

Maybe the harbinger was the fact that the bartender immediately thought my ID was fake (it wasn't). After all these years in the service industry, I should have known.

———————————

"Can I touch your hair?"

I stared at him from across the wood table. "What would your wife think of that?"

He gave me a sheepish look. "She probably wouldn't like it."

I continued staring, totally and completely taken aback by the question. We were several hours into the dinner, which up until this moment I thought had been a success. Business had been discussed; his next trip to DC put on my calendar; he had even mentioned flying me out to visit his Lockheed location. We had discussed which job posting would be the best for me to apply for and how he would bring attention to my application. A complete success . . . right up until that moment.

"Well then . . . would you like to come back to my hotel room and *just hold me*?" He leaned forward just slightly in his seat.

I leaned back, abruptly aware of how small the table was. "Uh . . . definitely not."

"You could just—"

"I think I should go." I stood up, almost tipping my chair over with the quick movement. "You can pay the bill," I added.

"Let me get you a taxi." He started to stand as well.

"No—"

"But—"

"I can get my own."

He followed me out. "Let me walk you to—"

"I'm fine, thanks," I said it louder this time, Uber already open on my phone. The door to the restaurant closed shut behind me. The nearest car was one minute away. *Thank god for rideshare apps. DC traffic might be hectic, but at least you could expect transport in under a minute.*

My heels clicked in a rapid beat across the cobblestones along the wharf as I moved toward the car. They were almost perfectly in time with the race of my heartbeat.

The director followed me, keeping several paces back.

I ignored him.

He wasn't a big man. And I was a tall girl. We only had to pass one alleyway. These were the thoughts that worry their way through a woman's mind in moments like this.

Thankfully, however, he kept a distance behind me until we reached the car door. There, he neared, opening his mouth as if to make one final plea, or perhaps—in recognition of his stupidity—an apology. I shut the door firmly and quickly, holding my breath. I was far more concerned about him trying to get in after me than slamming the door on his hand.

The driver looked back at me. I looked up at the driver.

"Columbia Heights."

"I know," The driver said.

"Right," I exhaled, laying my head back against the leather head-rest. I did not look out the window.

From that point on the drive was blessedly quiet, but within me a stormy panic slowly turned to a full-blown maelstrom.

Hell hath no fury like an angry anthropologist. And if you'll recall, this was not the first time something like this had happened to me. It was perhaps just the most uncouth example. At Spaceport America, while conducting fieldwork, I was asked out four times. This alone I don't see as an issue. Asking out someone you like, even someone you work with, is fine, in my opinion, but you have to understand that at one point the program director of one of Boeing's projects based out of the spaceport became so obsessively territorial in his interest in me (which, let's be very clear, I did not reciprocate) that he attempted to prevent me from conducting interviews with any of his male employees. To eventually do so, I had to wait until he was in the bathroom and then slip them my business card.

Once I married my second husband in 2019, I had multiple male colleagues in the space industry, who I frequently saw at space events and other group activities, inform me that we couldn't hang out because I was now married. As if our friendship had been one big ploy to eventually sabotage my engagement. In other instances, the reverse extreme occurred. Because I am polyamorous and try to support sex positivity, it was not uncommon for individuals to learn this and assume that meant I had no morals or sense of self-control.

The Uber dropped me off in front of my tiny studio apartment near Columbia Heights, and I slumped against my front door as it shut behind me.

I counted to ten and imagined jettisoning yet another man out an airlock. It didn't help one bit.

I tried again, unzipping my dress and taking several deep breaths. But deep breathing quickly turned to hyperventilation, and in a rage, I tore off one of my high heels and threw it at the wall.

My aim askew, it hit the bookshelf, and a used copy of M. T. Anderson's *Feed* hit the floor hard. An omen so blatant I could not ignore it.

"*We went to the moon to have fun, but the moon turned out to completely suck.*"

The first line of the novel was hard to forget.

It was written on the cover of the damn book.

"*We went . . .*

*to the moon*

*to have fun*

*but the moon*

*turned out*

*to completely*

*suck.*"

And I suddenly missed Florida with all of my being. Missed the openness of it. Despised the inescapable, claustrophobic hive of the city currently around me. I wanted . . . no, *needed* . . . to feel soil, sand, earth, and knew I would have to leave city lines to feel it. I craved the blanket of humidity and hallowed moss-covered ground. Felt the weighty absence of tidal ebbs and flows.

I tried to cling to a memory, in an act of psychological grounding, and sought out those most tellurian: mango picking during summers spent just north of Miami.

As children, barefoot and mosquito-bitten, we would eat stolen fruits half-ripe in the canopy of a banyan tree. I wish I could say juice dripped down our chins, but in all honesty, the fruit never seemed to swell to fullness, always staying in vegetal limbo. But we didn't know any better. We simply picked the ants off and enjoyed the earthly delights. And once finished, we rode our bikes, with full bellies, past other fruit and nut plantations nearby, waving gaily to migrant laborers.

Surely this strange fruit also offered strange foreshadowing.

---

The next morning, I told the women of CSF what happened at my meeting: "Should I report him? What if other women in the future don't have the confidence to say no to him?"

Their support was . . . restrained. Of course they would back whatever I wanted to do, *but* . . . and that "but" was monumental.

But I was on the job market!

But think of the scandal!

But maybe it wasn't worth the fallout?

But wasn't I interviewing for other positions? Didn't I just have dinner that week with someone from Momentus? And Boeing? I was still talking to Boeing, right? What about the Planetary Society? They adored me! Plus wasn't there someone else at Lockheed I had been speaking to? Someone in a different sector?

I had spoken in depth to Rob Chambers at Lockheed Martin. Rob was their director of business development and human spaceflight strategy. He shared my creative vision and specific interest in analog sites, and I had definitely seen a potential future working for him, but Rob, I realized, was based out of the same Lockheed location as the other director! How could I pursue working for one without running into the other?

And this is how harassment limits opportunities for victims. No matter how confident the victim of harassment is, or what type of harassment occurred, acts of harassment immediately place the victim in a difficult position. Do they report their harasser (if there is even a system to do so) and risk facing unpredictable consequences such as social stigma, lack of professional support, lack of repercussions, gaslighting and blame by the harasser, or legal repercussions, or do they not report the harasser and allow the individual to continue to get away with such behavior, perpetuating systems of intimidation and abuse? Not all places of employment have systems that allow for anonymous reporting, and not all employees—especially not young interns—have the emotional support systems, knowledge, or confidence to even understand how to report harassment and abuse.

Crestfallen, I let the issue go.

———————

During one of those last weeks at the Commercial Spaceflight Federation, I went to Colorado Springs, where I attended the annual Space Symposium—one of the largest space conferences in the world. While there, a friend in the industry who works closely with Lockheed Martin came up to me and said, "So remember that thing that guy said to you about 'holding him'?"

He giggled and continued, "I asked one of his former interns if she ever had someone ask her to 'hold them' and you'll never guess what she said . . ."

It seems that what had happened to me was not an isolated incident.

Beyond my own experiences of sexual harassment, something opened the floodgates in the years that followed my internship at the Commercial Spaceflight Federation. Quite a few incidents of sexual harassment came to light across the space industry, revealing long-standing systemic issues yet to be resolved. Systemic issues that were perhaps part of my own tipping point.

Those rose-colored glasses were well and truly shattered. The intoxicating fever dream that was human space exploration had dulled, and I was absolutely unsure what to do with the stars that had fallen from the sky.

Worn down and feeling absolutely disempowered, I went to a dinner in a conference hall that evening in Colorado Springs. Richard Branson, founder of Virgin Galactic and many other renowned companies, spoke. I sat between two of my closest friends in the industry around an ornately dressed table, and in the middle of Branson's speech, one of them reached over and grabbed my hand and squeezed it. She had tears of absolute devotion running down her face. Pure allegiance to the cause. Wholehearted and unbridled love.

And all of it was flowing out, purely because she got to see Branson, one of her idols, speak.

I squeezed her hand back and started crying, too. But not for the same reasons.

It wasn't just the sexual harassment. Or the confines of professionalism. The tightly woven black wool coat of a social scientist in a STEM world. It was all of it.

It was the fact that I, too, should have been reverently worshiping the colonial space capitalist idols, but knew in every fiber of my being that the future they were envisioning went against every decolonial thing I had learned as an anthropologist. That rather than sitting there at that table listening to Branson talk about the future of space tourism I should have been protesting it. And I didn't know how to do so, but I knew that someday soon I would.

# PART III

# A LETTER TO THOSE LEFT BEHIND

# 13

# AN ANTHROPOLOGIST'S CALL TO ARMS

*Woodford Reserve Distillery, Lexington, KY, 2019* CE

**In June 2019,** I got a job as a science writer at the American Institute of Physics. It was a wonderful place to work, and I can see why many work there for decades. Its structure is reminiscent of academia, its hierarchy respectful and equitable, and it's not often you can find a well-paid job as a full-time writer.

I liked being a journalist, which is what I essentially was, and it was strange how cathartic it was to let go of the idea of working in the space industry. To realize that I probably would never fit into the world of human space exploration unless a role was intentionally built for an ethicist (and such decolonial positions should be created with the intent to break down colonial structures and ensure sustainable space exploration). But what was far stranger was that the more determined I became to let go of my own cosmic dreams, the more space organizations suddenly wanted to listen to them.

In late summer 2019, I wrote "Lunar Imperialism (and How to Avoid It),"[1] and shortly after I received a call from Kris Kimel, the cofounder of Space Tango, a company that works directly with the ISS. Kris was creating a nonprofit dedicated to exploring issues surrounding human space exploration through nonindustry perspectives. In the process of doing so, he wanted to hear from poets, chefs, psychologists, teachers, clowns, actors, authors, and . . . anthropologists. With the help of the Planetary Society and several other sponsors, he was arranging an exclusive seminar in Lexington, Kentucky. Complete with an all-expenses-paid reception at the Woodford Reserve Distillery.

"You just gotta be there. I think your work is really incredible."

"My work?" Remember, at this point in time, I had published a handful of posts in academic blogs and a masters dissertation known to a select few.[2] Did I think I was developing a body of work? No. But Kris certainly did, and he *saw* something in it long before anyone else did.

Kris saw my research and me not as a web of questioning botanical tendrils, but as a fully formed forest, and he had every intention of cultivating and conserving such a forest. Even years later when, at times, our celestial goals have diverged (his have always been interstellar and mine remain Earthly) he listens with eagerness and an intense desire to learn from others. We work together to this day.

"Space anthropology! It's going to be at the forefront of this. And I want you to talk to this reporter while you're at the seminar . . ."

"A reporter?"

"Yes, her name is Leslie . . . and she wants to do a profile on you."[3]

To have this conversation with Kris, I had hidden in an unlit conference room at the American Institute of Physics. It's a little ridiculous looking back on it, but I used to sneak around to answer phone calls from space organizations, government agencies, and other magazines I was writing for. I was worried my employers would think I was too invested in my external work and it was distracting me from my job. I was still balancing two careers on my shoulders, still piling on extra work and keeping one foot in another world. And let's be clear, it totally was distracting me from my job! How could it not?

Most of my day was spent summarizing physics articles and writing press releases. So when NASA called, *Physics Today* had to wait.

The American Institute of Physics, as incredible a place to work as it was, always felt like a transitional zone for me. By that point, I knew that I eventually wanted to pursue a career in academia. I looked forward to teaching more than anything else, and what's more, to working in an environment that would encourage my creative and scholarly instincts.

"But . . . but . . . Kris . . ." I whispered, "my work, it's not always . . . sometimes I write about . . ."

I hesitated, and sighed.

". . . ethical issues . . . related to outer space."

And he responded with absolute conviction: "Well that's what we should be doing. We have to think about these things if we're going to really explore the stars. That's why we need people like you."

———————

That trip to Lexington was empowering. I was no longer working directly for the space industry, and the support that the community of similarly minded people offered me was like a hand grasped around mine, dragging me out of my shell. In the midst of space industry executives and commercial space entrepreneurs and poets and chefs and artists and, yes, clowns (no, there actually was someone there from Clowns Without Borders), I did not hold back one single ethical criticism of the space industry's interstellar goals.

For every technocratic, value-laden statement declared, I rebutted with culture and history and questions of class and race and sex. Arguments that I have revisited throughout this book. My protests were decolonial. They were historic. They were a self-liberation from the idea that I owed the space industry anything and that I had to believe in a long-standing scientific goal of human space exploration simply because humankind had been attempting it for an extensive period of history. A belief system, I realized, is not righteous just because a celebrity supports it or a president tweets about it or because the government declares it as "progress."

What made that seminar in Kentucky all the more worth it was that at the end of the first roundtable, the moderators asked a student from the University of Kentucky to come up and give her opinion on space exploration and its future. Without missing a beat, she began a beautiful monologue—to the surprise of the entire room—on how we should end human space exploration and prioritize issues on Earth. She was probably the same age I was when I began my journey as an outer space anthropologist, running down the basement corridors of the University of Florida's anthropology department.

Mat Kaplan, the Planetary Society's radio host for over twenty years, guided the seminar, and by happy coincidence, after the two day event was over he and I got stuck in the Lexington airport together. Both of our flights were delayed by several hours, and so we sat and spoke while we waited for them to take off. As we sat at the airport gate together, I asked him, heart in my hands, if I was . . . *too much* . . . at the seminar. Was I too bold? Too loud? Were my opinions too strong? My ideas too radical? To be clear, I think a lot of young professionals and women wonder such things.

And he looked me dead in the eyes and said, "No, I think it was really good, Savannah."

In the years following the seminar, space organizations—public and private—began to see the value in the anthropology of outer space. I received more and more invitations to write and speak about the topic, and I wrote two more invited articles in the following twelve months, and a third shortly after. I then spoke at the National Air and Space Museum, as well as two conferences.[4] The following year, the number of appearances I did doubled, and in 2022 that number tripled.

But, I still have those moments of insecurity. Just this past June I was invited to speak at a NASA workshop, and I remember saying afterward that I was consistently surprised they kept inviting me back. To which they laughed at me in response. A lot of that self-doubt stems not just from carving a path as a social scientist in a technocratic world or because of my age or sex but also because of what space exploration means to the individuals I work with. Space exploration isn't only a goal . . . it's a belief system. A historically ingrained, globally accepted belief system. And as flawed as it is—as colonial and imperialist and nationalist as it is, it is just as power-fully captivating.

Addictive, breathtaking, enthralling, tempting.

So vast and yawning is the poetry of space that some look into its depths and feel absolute fear. Others, awe.

And some, *desire*.

Insatiable, carnivorous desire.

Ravenous.

Covetous.

The type of worship that leaves behind billions on Earth.

And what to do about it?

Because *do* we must. Now I seek to reach beyond critique and raise a fist in the air with an answer grasped tightly between my fingers.

Shall we begin?

# 14

# THE CARETAKER'S DEMAND

*The Ministry for the Future, Zurich, 2025* CE

**First, we must make a demand.** The Caretaker's Demand.
The Caretaker's Demand, like I mentioned in the introduction of this book, asks us to wait, to care, to repair, and to conserve. It doesn't ask us to *never* go to outer space or to let go of those goals entirely. It seeks from its followers a reprioritization of values and a *patience*. A recognition that humans are not ready to settle outer space currently, and that motivations are presently and have historically been misplaced. It acknowledges that before we settle outer space, planet Earth and its global population must reach a point of stability with no tipping point in sight. So that nobody has to feel guilty for leaving anybody behind, and the population that chooses to stay, rather than go—when the time comes—is a population who has never imagined that the grass is greener and is making the decision to stay *themselves*.

This can, in part, be accomplished through the implementation of post-scarcity and Green New Deal political theory.

When I first heard about post-scarcity economic theory, in a guest lecture by Aaron Benanav, author of *Automation and the Future of Work*,[1] I remember after Benanav left the room I turned to the professor who hosted him and said, "Is this guy for real?" I was just so shocked that anyone could envision such a radically different future. A future in which all humans have their basic needs—food, water, housing—met, and in which labor and resources are communally distributed, cutting down the average workweek to a mere one or two days. This distribution of work would leave a tremendous amount of time for creative and intellectual pursuits.

At first, I was dumbfounded by the prospect, and the fact that Benanav had written a whole book outlining (in very economic terms) how to achieve the postcapitalist world.[2] I was even more shocked when I found out other political theorists had gone even further, conjuring up *multiple* potential utopian and dystopian versions of post-scarcity futures. Proposals that argue that capitalism *will end*, but that the form this ending takes is what is still up for configuration.[3] The more I started to think about it, the more I thought that such a radical future was kind of beautiful, and that the prospect only seemed so far-fetched because it was so progressive. I wasn't used to modern scholars developing the philosophical foundations of truly new worlds and suddenly felt all the more empathetic for great thinkers of ancient times—Aristotle, Pythagoras, Euclid, Archimedes, etc.—who paved the way for such things and likewise the postmodernist movement, which relies on such forms of social radicalism.

Now, let me be clear, I don't think we can achieve *utopia*. The word, stemming from Greek, written first in Latin, and coined by Thomas More in his satire by the same name, originally translated to "nowhere," but when shifted to English meant "good place" or "happy place."[4] It was a fictional city and remains a speculative ideal. But I think global *stability* might be achieved. This would be the point, as I mentioned before, at which socioeconomic-ecological catastrophic tipping points are few and far between.

However, I found myself repeatedly returning to Benanav's book—a book full of that dreaded enemy of mine . . . *mathematics*—and other books on the topic of post-scarcity.[5] I faced those arithmetic beasts down, thanked the gods I had passed statistics with high marks, and noticed that these scholars' research and mine had something in common. What if postcapitalist futures could help realize sustainable human space exploration?

Much of post-scarcity political theory is based on socialist ideas, and some post-scarcity economists have gone so far as to argue that such a future couldn't exist under any other type of government.[6] But we also see echoes of post-scarcity through the Green New Deal social movement introduced in the late 2010s. This political and scientific movement seeks to prioritize the climate crisis, restoration of the planet, and also many of the ideas held at the forefront of post-scarcity political theory.[7]

Both the Green New Deal and post-scarcity theory—however they might be accomplished (theorists from both camps describe recipes for action)—lay a solid foundation for a society without ecological tipping points in sight. One in which climate change has been stabilized, poverty and food insecurity neutralized, and most important for the purposes of this book, the human race is *ready* to go to outer space. It is a future in which the humans that do go to space will do so with no intent to replace or acquire, because they will have no need for such things. They will have been raised not on the rhetoric of capitalism, colonialism, and territorialism but instead on preservation, maintenance, and conservation.

Ironically, the design and implementation of such futures can be sped up with the help of disastrous events—or *pressure*. Dystopian moments have a way of pushing social changes forward at tremendous speed. For example, I've mentioned the COVID-19 pandemic before, and we can return to it as an example again.

Consider the speed at which certain cultural laws shifted as a result of the pandemic. Take laws surrounding alcohol sale and consumption, for example. Within days of shelter-in-place ordinances and the closing of nonessential businesses, many states rescinded laws that previously restricted restaurants from selling alcohol to go.[8] Not only that, but rules surrounding online liquor sales and direct sales with breweries and wineries were also loosened. In most jurisdictions, the sale of alcohol was deemed "essential," with the exception of the state of Pennsylvania, which closed all state-run liquor stores.

The reason for these policy changes varied across the board. Some claimed it was economic; some pointed to the social and mental repercussions of removing easy access to a substance that many individuals in the world are addicted to (to varying degrees). No matter the reason, the United States—a country with notoriously strict laws surrounding the sale and service of alcoholic beverages and a fascinating history of prohibition—suddenly relaxed the majority of their restrictions.

This might seem like a silly example to draw upon, but it is an astonishingly accurate reminder of how easily long-standing beliefs and values can shift when faced with catastrophe. The COVID-19 pandemic offers plenty of other examples as well: Quarantine restrictions. Social distancing indicators. Rapid technical and scientific development. And I have no doubt we will see similar rapid social and political changes when other viruses, overpopulation, climate change, or nuclear warfare reach their tipping points—or a moment of collapse. Climax. Crescendo. True planetary-scale ecological holocaust. Immense *pressure*.

The thing is, we don't have to wait for disaster to strike to convince the human population to focus their priorities Earthward.

*Pressure* can come from social movements. Political shifts stemming not from disaster but passion in a cause, political reprioritization, protest, and *multigenerational* focus.

This idea of thinking generationally plays a large role in this call to arms and is very inspired by science-fiction works that have explored the aftermath of space colonization on the children of colonizers. Not to mention real multigenerational policies being put in place, such as those created after the Mauna Kea protests, which give Native Hawaiian cultural experts voting seats on a new governing body and take into consideration environmental factors and the multigenerational impact of astronomical construction.

Novels by Kim Stanley Robinson, Orson Scott Card, and Joe Haldeman have described the multigenerational repercussions that can stem from what seems like an expertly planned colonization effort.[9] Books like these can make us realize that it's difficult to plan human space exploration past one generation. Generation one might be the best of the best in terms of training and intelligence, but you can only control genetics so far, and you can't control *thought*. (Though many have tried.)

Generations two, three, four, and so on become harder to plan for as a spacefaring civilization evolves, and they also become harder to control socially or biologically. In the end, one has to ask, "OK, what if we convince generation one to adopt the Caretaker's Demand, which asks humanity not to leave planet Earth but focus on its care and maintenance? How do we maintain that demand across generations two, three, four?" Especially when this book has literally just demonstrated how volatile even the best socio-technical political plans are when bent to the socio-cultural will of the public. Maintaining multigenerational policies, let alone beliefs and values, is a complicated cultural process and hard to manage synthetically. So how would something like the Caretaker's Demand be institutionalized without institutionalizing a totalitarian state?

The thing is, Robinson also proposed a different idea, in another one of his novels, *The Ministry for the Future*.[10] The book follows a governmental structure developed during a time of climate crisis, which advocates for future generations based on decisions happening in the present. The rights of these people—not yet born—are considered with equal weight to those existing in the present. This is the mindset—if not the actual institution—we need to ensure that the Caretaker's Demand is sustained. An enforcement not of multigenerational values (this is the part that is totalitarian and is often ineffective), but of *recognition* of multigenerational impact. In other words, the creation of a governing body that would require that techno-scientific plans, especially those with a global reach, take into consideration multigenerational effects and multigenerational voices.

Much like those cultures that carry with them voices of their ancestors to guide their present actions, we must seek to carry with us the voices of our descendants before they are born.

Other voices must be included as well.

The quickest way to begin this celestial revolution is to include social scientists in techno-scientific development, and to ensure the diversity of congressional testimony.

Such a request might seem simple, especially considering the push in recent decades toward inclusion, diversity, and interdisciplinary programs. But when it comes to science and technology projects, as well as academic programs, the term "interdisciplinary" is often interpreted either lightly or taken advantage of to achieve other goals. By interpreted lightly, I mean that an engineering department, for example, will say that they are "interdisciplinary" because they've gone wild and crazy and started mingling mechanical engineers and electrical engineers. The thought of drawing from anthropologists, sociologists, historians, or other social scientists (let alone artists, poets, storytellers, etc.) remains the distant daydream of those under-funded and outcast.

Interdisciplinarity needs to go beyond the inclusion of subfields within a field. Instead, social scientists, ethicists, and historians need to be included in major techno-scientific projects to help prevent negative feedback cycles and the continuation of harmful patterns of imperialism, territorialism, colonialism, racism, sexism, and classism. Accomplishing this can be as simple as hiring individuals in these disciplines as project consultants, but the government could go so far as to make the inclusion of these individuals on projects mandatory when it comes to the approval of major techno-scientific ventures (like human space exploration). Consider the role sort of like an ethical health inspector. In addition, Congress might consider a revision of their selection process for congressional witnesses, or the institution of a similar staff position dedicated solely to the task of witness selection.

Beyond that, human space organizations also need to work to involve the voices that have *never* been included in conversations about outer space. For example, those from the disability community, who have already proven through programs like AstroAccess and the JustSpace Alliance that they can thrive in space. Indigenous populations should also be included, both because they deserve a position in this conversation as equals (not just because rockets are being launched from their land) and because they possess valuable knowledge they can share if they deem the rest of the human population worthy of knowing it. Indigenous populations know how to survive in extreme conditions, how to live with the land and not on it, how to utilize in situ resources, how to mediate conflict under extreme conditions and in confinement, and how to understand territories and land ownership (or the lack thereof) in alternative ways.

We can also draw such knowledge from other underestimated populations who work in extreme, isolated, or high-stress environments, such as industrial laborers, seafarers, islanders, hospitality industry workers, and more.

What is crucial throughout this process of inclusion, though, is that these individuals are valued and give continuous consent as they share their knowledge. They must understand the roles they play in conversations about human space exploration and be treated as collaborators—not resources. Inclusion is a balancing act. New voices must be involved in the development of techno-scientific projects and alternative knowledge structures must be acknowledged, but the individuals must be included in the conversation from start to finish.

To realize the Caretaker's Demand and ensure an equitable human future in outer space, much more must occur than a change in political structure, policy, or the establishment of a new institution.

Reprioritization has to occur first *at the personal level* and then the political level. Because prioritization is how promises and prophecies become reality.

I am making promises.

Now *you* need to prioritize them.

Your voice matters in conversations about outer space too.

# EPILOGUE

# EARTHWARD AUGURIES AND ACTIVISMS

*Palm Beach Gardens, FL, 2011* CE

**When I was fifteen,** I skipped class, went to the high school library, and found a book printed in the 1970s or '80s titled *Seven Steps to Conquering the Universe*. I've struggled to track down a copy of it since. Maybe I'm misremembering how many steps there were in the process. For all I know, I could have the title completely wrong. But what I remember so clearly about that book was that it argued that before we could even think about "conquering" outer space, humans should "conquer" oceans, focusing construction efforts on building vast underwater cities out of coral and harvesting hydrothermal energy.

Just over a decade after I found that book, I went to a temporary museum exhibit in Washington, DC, located right on the Washington Mall in the Arts and Industries Building. There, the Smithsonian was showcasing a display of futurism—from both a historic perspective and a speculative perspective, and in the present day. And not far from a prototype model of Virgin Hyperloop, they had this little exhibit that wasn't receiving a ton of attention from much of anyone. Most of the other onlookers were busy snapping photographs with the model of the giant train pod and interacting with various forms of artificial intelligence throughout the museum, but I stood, hands in the pockets of my jeans, and stared at this tiny display they had of cities built across the Earth's oceans. It happened to be set really close to a version of the first video game developed by Arctic peoples, *Never Alone* (*Kisima Ingitchuna*), a survival game that draws upon Indigenous storytelling.

As I looked down at cities inspired by the Golden Age of science fiction and lingering eco–New Wave values built across the surface of a still sea made of epoxy, I couldn't help but think of that book from my childhood. I knew the idea of cities built across the ocean was simply part of an era of design motivated by Buckminster Fuller and other forms of eco-communist architectural philosophy. I just wondered what happened to such reveries. Like every other techno-scientific idea, surely such Earthward imaginings were also part of sociopolitical cycles. So at what point did this dream lose favor? When were eco-friendly oceanic cities smothered by their capitalist spacefaring twin? And how could such ideas be revived?

Was my desire to revive them wrong?

It was 2022, and by this point I no longer wrestled with the judgment of my peers in the commercial space industry. I was, however, in Washington, DC, that weekend to discuss with some of my closest friends the prospect of writing this book. Out of curiosity, more than anything else, I wanted to hear their thoughts on a publication dedicated to decolonizing space exploration. So at the birthday party of a friend in the industry, hosted at a whiskey bar near Chinatown, I started asking friends and strangers what they thought of a book that questioned whether it was worth it to go to space at all. To my complete surprise, the response was overwhelmingly positive. Sure, they still wanted to go to space. And yes, they did idolize it. But something wasn't right about the way things were being done.

But how did such activism fit into anthropology? How could one both be a neutral observer and fight for a cause? The prospect went against a lot of my basic training. I had been taught about bias and to be up front with it, but knew I should maintain neutrality as much as possible. As it turns out, though, activism fit into a lot of other anthropologists' training,[1] and coming to understand this was the final shift in my transformation. A shedding of skin. A blossoming. A big bang. Use whatever metaphor suits your fancy. For my field of study—the anthropology of outer space—activism looked like applying decolonial methods and anticapitalist political theories to my research results. It looked like influencing law and policy through qualitative research. Some call it *engaged STS*, or science and technology studies. In other words, a *response* to what I saw—not a betrayal—along with an offering of possible solutions to long-standing systemic issues. I carefully designed new crystal clear glasses to replace the rose-colored ones.

For some anthropologists and social scientists, activism can mean participating in protests and helping to protect certain rights or values. Activism is a response and an application of knowledge gained. Sometimes in big ways, like trying to realize radical new futures, and sometimes in small ways, like installing a consulting hour at a clinic.

I stared down at tiny model white-walled cities strewn across a plastic sea, lost in an ocean of Earthward auguries and activisms until I felt a tug on my sleeve.

"Sav, they have an AI that will choose a New Year's resolution for you!" One of my friends in the industry waved a little colored card in front of me. She grinned up at me.

"Look, mine says . . . 'I will start believing that insects are ghosts wearing tiny costumes!'"

I laughed and looped my arm in hers, forgetting about everything except for what was most important: the here, the now, and the Earthly, "OK, I have to see this . . ."

# ACKNOWLEDGMENTS

**A**cknowledgments are like trying to say thank you to a bunch of people at once. Far more than thank you, actually. Too often, acknowledgments become a list of loved ones, when what you really want to say is "My god, you didn't know it, but that Wednesday in September when we talked about chapter 10, you stopped me from giving up" or "I am so glad you convinced me not to include that." And there's this frantic worry I think most writers probably have that you'll leave someone out. Someone so deserving of recognition that a week after publication you'll wake in a cold sweat and go "Oh no! I forgot about Greg!"

I don't actually know a Greg.

At least I don't think I do. But if a Greg knows me, then I thank you, Greg.

I do however know some incredible professors and staff at the Virginia Tech Department of Science and Technology Studies who have supported me throughout this project and, more important, have consistently supported my out-of-the-box, wacky, creative ideas. You have been nothing but appreciative of my boundary pushing, and I am forever grateful for that. Thank you, Lee, Matt, Saul, Daniel, Ashley Shew, Christine, Fabien, Jim, Gary, Carol-Sue, Ashley Snider, and Matthew G. And thank you, Matthew V., for being there through every creative ebb and flow. I would also like to thank the advisers I've had at other academic stages in my life. Without you, I would not be where I am today. Thank you, Jeeva, Victor, Aaron, and Richard. Among you are many space anthropologists and space sociologists who have been incredible in their support as we work together to study the celestial and the Earthly. These tiny botanical networks of academic knowledge and companionship birth lush forests of knowledge. May they continue to flourish.

As ironic as it might sound, there are also *so many* individuals in the space industry who have supported me on this journey, on both a deeply personal level and a scholarly one. I want to acknowledge every single one of you who seeks a decolonial, cosmic world. Those who have stood up for women in space, for the voices left out of conversations, and for me just because we're friends! I would especially like to acknowledge the work being done by AstroAccess and the JustSpace Alliance. As well as the support and friendship of Kris, Alex C., Joseph, Cody, Cameo, Therese, Luc, Marchel, Billy, Martin, Renatta, Victoria, a special space economist, a couple of astronauts, Kim Stanley Robinson and his publicist (whose name I really can't remember, I'm sorry, but I fangirled super hard when you all emailed me, and it kept the inspiration flowing), Nick and the Interstellar Foundation, Greg S., the folks at NASA who keep inviting me back, and so many others. I would also like to thank the Spaceport America team for hosting me, and Eric Stallmer for being an incredible boss during my time at CSF.

And, of course, where would I be without my parents, my sister, and the childhood friends who helped me shatter my rose-colored glasses? I really could not have done this without you. You sat and listened to me struggle to work out paragraphs and arguments and evidence even when you had no idea what I was talking about. For that I am forever grateful.

In the same vein I should acknowledge my friends, fellow bartenders, and the other free spirits that have kept me grounded here on Earth as I researched the cosmic and surreal. Thank you Lo, Kat, Travis, Justin, Miles, Madigan, Sam, Paul, Sarah, Lois, Lyndon, the entire staff of Eastern Divide and Rivermill, and all my fellow graduate students.

Lastly, I want to thank my editor Jerry and the entire team at Chicago Review Press, as well as my agent, James McGowan, and the Bookends team. James, you put the *power* in *publishing power couple.* Kismet brought us together, and dedication, commitment, and faith have kept us hard at work. Thank you for everything.

# NOTES

## Introduction: The Protest

1. Alex Knapp, "Understanding the Thirty Meter Telescope Controversy," *Forbes*, June 12, 2015, https://www.forbes.com/sites/alexknapp/2015/06/12 /understanding-the-thirty-meter-telescope-controversy/; Tim Fernholz and Peter D'Auria, "How a Handful of South American Protestors Took Europe's Space Program Hostage," *Quartz*, April 17, 2017, https://qz.com/960817 /how-a-handful-of-south-american-protestors-in -french-guiana-took-arianespace-and-europes-space-program-hostage/.

2. Ursula K. Le Guin, "The Ones Who Walk Away from Omelas," *New Dimensions* 3 (1973).

3. Dominik Reuter, "600 People Have Reserved $250,000 Tickets to Fly to Space with Virgin Galactic, Including Celebrities like Tom Hanks, Leonardo Dicaprio, Justin Bieber, and Lady Gaga," *Business Insider India*, July 10, 2021, https://www.businessinsider.in/tech/news/600-people-have-reserved -250000-tickets-to-fly-to-space-with-virgin-galactic-including-celebrities -like-tom-hanks-leonardo-dicaprio-justin-bieber-and-lady-gaga/articleshow /84281007.cms.

4. The following three paragraphs are adapted from the author's publication "The Death of Outer Space Dreams: Hard Decisions and a War of Utopian Demands," *Geek Anthropologist*, April 22, 2021, https://thegeekanthropologist .com/2021/04/22/the-death-of-outer-space-dreams-hard-decisions-and-a -war-of-utopian-demands/.

5. Charlene Carruthers, "Hearing Assata Shakur's Call," *Women's Studies Quarterly* 46, no. 3/4 (2018): 222–225; Assata Shakur, *Assata: An Autobiography* (Chicago: Chicago Review Press, 1999).

# 1. Past the Potato and into the Future

1.  Valerie Olson, "American Extreme: An Ethnography of Astronautical Visions and Ecologies" (PhD diss., Rice University, 2010).

2.  Stefan Helmreich, *Alien Ocean: Anthropological Voyages in Microbial Seas* (Berkeley: University of California Press, 2009).

3.  Janet Vertesi, *Seeing Like a Rover: Images and Interaction on the Mars Exploration Rover Mission* (Chicago: University of Chicago Press, 2015).

4.  James Gurney, *Dinotopia, a Land Apart from Time: 20th Anniversary Edition* (Mineola: Calla Editions, 2011).

5.  Savannah Mandel, "On Refrigerators and Rage: Secrecy and Pacification in the Florida Restaurant Industry," *Journal of Contemporary Ethnography* 52, no. 1 (2021).

6.  Clifford Geertz, "Deep Play: Notes on the Balinese Cockfight," *Daedalus* 101, no. 1 (1972): 1–37.

7.  Magoroh Maruyama and Arthur Harkins, *Cultures Beyond Earth: The Role of Anthropology in Outer Space* (New York: Vintage House, 1975); Peter Redfield, *Space in the Tropics: From Convicts to Rockets in French Guiana* (Berkeley: University of California Press, 2000); Michael Oman-Reagan, "Anthropologists in Outer Space," *Wanderers*, February 8, 2016, https://www.sapiens.org/culture/anthropologists-in-outer-space/.

8.  "Diary of a Space Zucchini" was written by astronaut Don Pettit and commented on by another space anthropologist, Debbora Battaglia. Debbora Battaglia, "Diary of a Space Zucchini: Ventriloquizing the Future in Outer Space," *Platypus*, July 14, 2014, https://blog.castac.org/2014/07/diary-of-a-space-zucchini-ventriloquizing-the-future-in-outer-space/.

9.  James McGowan, *Goodnight Oppy*, (New York: Astra Young Readers, 2021); James Vincent, "'Goodnight Earth. Goodnight Humanity': China's Jade Rabbit Rover Tweets Its Own Death," *Independent*, February 3, 2014, https://www.independent.co.uk/tech/goodnight-earth-goodnight-humanity-china-s-jade-rabbit-rover-tweets-its-own-death-9103864.html; Michael Greshko, "China's 'Jade Rabbit' Moon Rover Declared Dead," *National Geographic*, August 4, 2016, https://www.nationalgeographic.com/science/article/china-moon-rover-jade-rabbit-dead-yutu-space-science.

10. Lisa Messeri, "Extra-Terra Incognita: Martian Maps in the Digital Age," *Social Studies of Science* 47, no. 1 (2016): 75–94.

11. Valerie Olson, *Into the Extreme: U.S. Environmental Systems and Politics Beyond Earth* (Minneapolis: University of Minnesota Press, 2018).

12. Debbora Battaglia, *E.T. Culture: Anthropology in Outerspaces*, (Durham: Duke University Press, 2006).

## 2. The Crossroads

1. David Jeevendrampillai and Aaron Parkhurst, "Making a Martian Home: Finding Humans on Mars Through Utopian Architecture," *Home Cultures* (2021): 1751–7427; Timothy Carroll and Aaron Parkhurst, "Being, Being Human, Becoming Beyond Human," in *Lineages and Advancements in Material Culture Studies* (London: Routledge, 2020); Aaron Parkhurst and David Jeevendrampillai, "Towards an Anthropology of Gravity: Emotion and Embodiment in Microgravity Environments," *Emotion, Space and Society* 35 (2020).

2. Jeevendrampillai and Parkhurst, "Making a Martian Home."

3. "About," Ethno-ISS, accessed September 14, 2022, https://ethnoiss.space/.

4. European Commission, "ETHNO-ISS: An Ethnography of an Extra-terrestrial Society: The International Space Station," September 4, 2019, https://cordis.europa.eu/project/id/833135.

5. Jo Aiken and Angela Ramer, "From the Space Station to the Sofa: Scales of Isolation at Work," *Ethnographic Praxis in Industry Conference Proceedings*, October 2020, https://doi.org/10.1111/epic.12044; Victor Buchli, "Extra-terrestrial Methods: Towards an Ethnography of the ISS," in T. Carroll et al. *Lineages and Advancements in Material Culture Studies: Perspective from UCL Anthropology* (London: Bloomsbury, forthcoming); Victor Buchli, "Low Earth Orbit: A Speculative Ethnographer's Guide," in *Anti-Atlas: Towards a Critical Area Studies*, ed. W. Bracewell et al. (London: UCL Press, forthcoming); David Jeevendrampillai and Aaron Parkhurst, "Towards an Anthropology of Gravity: Emotion and Embodiment in Microgravity Environments," *Emotion, Space and Society* 35 (2020).

6. Robert Emerson, Rachel Fretz, and Linda Shaw, *Writing Ethnographic Fieldnotes* (Chicago: University of Chicago Press, 1995); John Levi Martin, *Thinking Through Methods: A Social Science Primer* (Chicago: University of Chicago Press, 2017).

7. Olson, "American Extreme," 16–17.

8. Olson, "American Extreme."

9. David Valentine, "Exit Strategy: Profit, Cosmology, and the Future of Humans in Space," *Anthropological Quarterly* 84, no. 4 (2012): 1045–1067.

10. Teun Zuiderent-Jerak, "Embodied Interventions—Interventions on Bodies: Experiments in Practices of Science and Technology Studies and Hemophilia Care," *Science, Technology & Human Values* 35, no. 5 (2010): 677-710.

11. Nicholas Schmidle, *Test Gods: Virgin Galactic and the Making of the Modern Astronaut* (New York: Henry Holt and Co., 2021).

12. Adam Rogers, "Space Tourism Isn't Worth Dying For," *Wired*, October 31, 2014, https://www.wired.com/2014/10/virgin-galactic-boondoggle/.

13. Ingrid Burrington, "New Mexico's Sad Bet on Space Exploration," *Atlantic*, March 2, 2018, https://www.theatlantic.com/technology /archive/2018/03/new-mexicos-sad-bet-on-space-exploration/554243/.

14. Associated Press, "Former New Mexico Spaceport CFO Alleges Fraud, Retaliation," *US News & World Report*, January 5, 2022, https://www .usnews.com/news/best-states/new-mexico/articles/2022-01-05/former -new-mexico-spaceport-cfo-alleges-fraud-retaliation; Chelsea Gould, "Report Finds that Former Spaceport America Director Violated State Laws (Report)," Space.com, December 4, 2020, https://www.space.com/spaceport -america-former-director-investigation.

15. Truth or Consequences, directed by Hannah Jayanti (Sentient.Art.Film, 2020).

## 3. On Faith and Sacrifice

1. Doug Messier, "A Brief History of Spaceport America," *Parabolic Arc*, August 19, 2019, http://www.parabolicarc.com/2019/08/19/a-brief-history-of -spaceport-america/ (site discontinued).

2. Jeff Foust, "Virgin Galactic and the Future of Commercial Spaceflight," Space.com, May 23, 2005, https://www.space.com/1110-virgin-galactic -future-commercial-spaceflight.html.

3. Doug Messier, "NM's Latest Plan to Make Money from Spaceport America: Stop Treating SpaceShipTwo Passengers as Freight," *Parabolic Arc*, January 27, 2022, http://www.parabolicarc.com/2022/01/27/nm-plans-money -spaceport-america-involves-longer-treating-spaceshiptwo-passengers -freight/ (site discontinued).

4. Kate Bieri, "How Much Has Spaceport America Cost Taxpayers?" KVIA .com, July 9, 2021, https://kvia.com/news/new-mexico/2021/07/09/how -much-has-spaceport-building-cost-dona-ana-county-taxpayers/; Messier, "NM's Latest Plan to Make Money from Spaceport America."

5.  Alice Goffman, *On the Run: Fugitive Life in an American City* (Chicago: University of Chicago Press, 2014).

6.  Philippe Bourgois and Jeffrey Schonberg, *Righteous Dopefiend* (Berkeley: University of California Press, 2009).

7.  Sections of the next three pages have been adapted from the author's dissertation, "The Astrophile's Mite: Risk and Sacrifice in the Commercial Space Industry," (MSc diss., University College London, 2018).

8.  Maggie Grimason, "Spaceport America: New Mexico's Protracted Hope for a Revived Economy," *Wire Science*, May 8, 2018, https://science.thewire .in/aerospace/spaceport-america-new-mexicos-protracted-hope-for-a -revived-economy/.

## 4. Asteroids and Access

1.  Savannah Mandel, "The Elysium Effect: Space Law and Commercial Space Disparities," *Geek Anthropologist*, July 10, 2018, https:// thegeekanthropologist.com/2018/07/10/the-elysium-effect-thoughts-on -that-future-we-dont-want-to-happen/.

2.  US Congress, Commercial Space Launch Competitiveness Act, H.R.2262, 114th Cong. (2015–2016), https://www.congress.gov/bill/114th-congress /house-bill/2262/text; US Congress, Commercial Space Launch Amendments Act, H.R. 5382, 108th Cong. (2003–2004), https://www.congress.gov/bill /108th-congress/house-bill/5382/text; US Congress, Commercial Space Launch Act, H.R. 3942, 98th Congress (1983–1984), https://www.congress .gov/bill/98th-congress/house-bill/3942/text.

3.  US Congress, Commercial Space Launch Competitiveness Act, section 51303.

4.  Jolene Creighton, "History's Future in Space Depends on Asteroid Mining." *Futurism*, June 23, 2016, https://futurism.com/humanitys-future-in-space -depends-on-asteroid-mining; Jack Heise, "Space, the Final Frontier of Enterprise: Incentivizing Asteroid Mining Under a Revised International Framework," *Michigan Journal of International Law* 40, no. 1 (2018): 189; Andrew Rosenblum, "How to Mine an Asteroid," *Popular Science*, December 17, 2011, https://www.popsci.com/science/article/2011-11/how-mine -asteroid/; Jamie Carter, "Are We Finally on the Cusp of Commercial Asteroid Mining?" *Sky & Telescope*, February 27, 2023, https:// skyandtelescope.org/astronomy-news/are-we-finally-on-the-cusp-of -commercial-asteroid-mining/; J. A. Dallas, S  Raval, J. P. Alvarez Gaitan,

S. Saydam, and A. G. Dempster, "Mining Beyond Earth for Sustainable Development: Will Humanity Benefit from Resource Extraction in Outer Space?" *Acta Astronautica* 167 (2020): 181–188.

5. Atossa Araxia Abrahamian, "How the Asteroid-Mining Bubble Burst," *MIT Review*, June 26, 2019, https://www.technologyreview.com/2019/06/26 /134510/asteroid-mining-bubble-burst-history/; Bruce Dorminey, "Does Commercial Asteroid Mining Still Have a Future?," *Forbes*, August 31, 2021, https://www.forbes.com/sites/brucedorminey/2021/08/31/does-commercial -asteroid-mining-still-have-a-future/.

6. Aristophanes, *The Frogs*, in *The Complete Greek Drama: All the Extant Tragedies of Aeschylus, Sophocles and Euripides, and the Comedies of Aristophanes and Menander, in a Variety of Translations*, ed. Whitney J. Oates and Eugene O'Neill (New York: Random House, 1938) 449.

## 5. A Trip Down the River Styx

1. Dwight D. Eisenhower, "Address Before the 15th General Assembly of the United Nations," September 22, 1960, US Department of State Archived Content, https://2009-2017.state.gov/p/io/potusunga/207330.htm.

2. United Nations, "Treaty on Principles Governing the Activities of States in the Exploration and Use of Outer Space, Including the Moon and Other Celestial Bodies," resolution 2222, opened for signature January 27, 1967, (XXI), https://www.unoosa.org/oosa/en/ourwork/spacelaw/treaties.html.

3. David Callahan and Fred Greenstein, "The Reluctant Racer: Eisenhower and U.S. Space Policy," in *Spaceflight and the Myth of Presidential Leadership*, ed. Roger Lanius and Howard McCurdy, (University of Illinois Press, 1997).

4. United Nations, "Exploration and Use of Outer Space," article I.

5. United Nations, "Exploration and Use of Outer Space."

6. United Nations, "Agreement Governing the Activities of States on the Moon and Other Celestial Bodies," resolution 34/68, opened for signature December 18, 1979, https://www.unoosa.org/oosa/en/ourwork/spacelaw /treaties.html.

7. United Nations, "Activities of States on the Moon," article IV.

8. United Nations, "Exploration and Use of Outer Space," article I.

9. United Nations, "Activities of States on the Moon," article VI.

10. US Congress, Commercial Space Launch Competitiveness Act, section 51303.

11.  Jesse Dunietz, "Floating Treasure: Space Law Needs to Catch Up with Asteroid Mining," *Scientific American*, August 28, 2017, https://www .scientificamerican.com/article/floating-treasure-space-law-needs-to-catch -up-with-asteroid-mining/.

12.  US Congress, Commercial Space Launch Competitiveness Act,  section 115.

13.  Feyisola Ruth Ishola, Oluwabusola Fadipe, and Olaoluwa Colin Taiwo, "Legal Enforceability of International Space Laws: An Appraisal of 1967 Outer Space Treaty," *New Space* 9, no. 1 (2021): 33–37.

14.  Ishola, Fadipe, and Taiwo, "Legal Enforceability of International Space Laws," 34.

15.  Homer, *The Odyssey,* trans. A.T. Murray (Cambridge: Harvard University Press, 1919) , 4.80.565.

16.  Hesiod, *Works and Days,* trans. by H. G. Evelyn-White, in Loeb Classical Library, vol. 57 (London: William Heinemann, 1914), 15; Plutarch, *The Parallel Lives*, trans. by Bernadotte Perrin, in Loeb Classical Library, vol. 8 (Cambridge: Harvard University Press, 1919), 9.

17.  Creighton, "History's Future in Space Depends on Asteroid Mining," https:// futurism.com/humanitys-future-in-space-depends-on-asteroid-mining; Alex Gilbert, "Mining in Space is Coming," *Milken Institute Review*, April 26, 2021, https://www.milkenreview.org/articles/mining-in-space-is-coming.

18.  Homer, *Odyssey*, 4.80.561–589.

19.  Sophie Goguichvili, Alan Linenberger, Amber Gillette, et al., "The Global Legal Landscape of Space: Who Writes the Rules on the Final Frontier?," Wilson Center, October 1, 2021, https://www.wilsoncenter.org/article /global-legal-landscape-space-who-writes-rules-final-frontier.

20.  This term was first proposed in the author's article "The Elysium Effect: Space Law and Commercial Space Disparities," however it is of no relation to an article of the same name published on Space.com in 2020 by Rich Tumlinson. The author accused Tumlinson of plagiarism in 2021 by contacting the editor-in-chief of Space.com. Tumlinson denied the accusation on the grounds that he "invented the term in an email in 2014," and the article remains online.

21.  Virgil, *Aeneid*. trans. Fairclough, in Loeb Classical Library, vols. 63 & 64 (Cambridge: Harvard University Press, 1916), 5.724.

22.  Goguichvili, Linenberger, Gillette, et al.., "Global Legal Landscape of Space," https://www.wilsoncenter.org/article/global-legal-landscape-space-who -writes-rules-final-frontier.

23.  Dorminey, "Does Commercial Asteroid Mining Still Have a Future?,"
     https://www.forbes.com/sites/brucedorminey/2021/08/31/does-commercial
     -asteroid-mining-still-have-a-future/.

## 6. Six Decades of Space Protests

1.   Elias Canneti, *Crowds and Power* (London: Victor Gollancz, 1962).

2.   Savannah Mandel, "Lunar Imperialism (and How to Avoid It)," *Association
     for Political and Legal Anthropology*, July 12, 2019, https://www.anthropology
     -news.org/articles/lunar-imperialism-and-how-to-avoid-it/.

3.   Neil M. Maher, *Apollo During the Age of Aquarius* (Cambridge: Harvard
     University Press, 2019), 11–12.

4.   "NASA Chief Briefs Abernathy After Protest at Cape," United Press
     International, July 16, 1969, https://www.upi.com/Archives/1969/07/16
     /NASA-chief-briefs-Abernathy-after-protest-at-Cape/7371558396299/; Eric
     Niiler, "Why Civil Rights Activists Protested the Moon Landing," History,
     July 11, 2019, https://www.history.com/news/apollo-11-moon-landing
     -launch-protests.

5.   Gil Scott-Heron, "Whitey on the Moon," track 9 on *Small Talk at 125th and
     Lenox* (Flying Dutchman Records, 1970).

6.   Simeon Booker, "Moon Probe Laudable—but Blacks Need Help," *Jet*, August
     1, 1969, 10; Thomas A. Johnson, "Blacks and Apollo: Most Couldn't Have
     Cared Less," *New York Times,* July 27, 1969.

7.   David Miguel Molina and P. J. Blount, "Bringing the Moon to Mankind: The
     Civil Rights Narrative and the Space Age," in *NASA and the Long Civil Rights
     Movement* (Gainesville: University Press of Florida, 2019), 51.

8.   "Americans Favour End to Moon Race," *Times*, April 9, 1963.

9.   Amitai Etzioni, *The Moon-Doggle: Domestic and International Implications of
     the Space Race* (New York: Doubleday, 1964).

10.  Niiler, "Why Civil Rights Activists Protested the Moon Landing"; Sonia Rao,
     "Why 'First Man' Prominently Features Gil Scott-Heron's Spoken-Word
     Poem 'Whitey on the Moon,'" *Washington Post*, October 13, 2018, https://
     www.washingtonpost.com/arts-entertainment/2018/10/13/why-first-man
     -prominently-features-gil-scott-herons-spoken-word-poem-whitey-moon/;
     Jane C. Hu, "First Man Offers a Helpful Reminder That a Lot of Americans
     Didn't Want to Go to the Moon," *Quartz*, October 25, 2018, https://qz.com
     /1432303/first-man-shows-that-many-americans-opposed-nasas-moon
     -mission/.

11. Jane Young, "Pity the Indians of Outer Space: Native American Views on the Space Program," *Western Folklore* 46, no. 4 (October 1987): 273.

12. James Andrews and Asif Siddiqi, eds. *Into the Cosmos: Space Exploration and Soviet Culture* (Pittsburgh: University of Pittsburgh Press, 2011); Asif Siddiqi, *The Red Rockets' Glare: Spaceflight and the Soviet Imagination, 1857–1957* (Cambridge: Cambridge University Press, 2010), 5.

13. Space anthropologist Debbora Battaglia wrote an article on this event. See Debbora Battaglia, "Arresting Hospitality: The Case of the 'Handshake in Space,'" *Journal of the Royal Anthropological Institute* 18 (2012): S76–S89.

14. For example, Bernie Sanders tweeted "What if instead of helping billionaires fund their space hobby, we invested in working people here on Earth?" on November 29, 2021: https://twitter.com/SenSanders/status/14653719022 39719427.

15. Sara Kahanamoku et al. "A Native Hawaiian–Led Summary of the Current Impact of Constructing the Thirty Meter Telescope on Maunakea," white paper submitted to the National Academy of Sciences Decadal Survey on Astronomy and Astrophysics panel on the State of the Profession and Societal Impacts, January 3, 2020.

16. Knapp, "Thirty Meter Telescope Controversy," https://www.forbes.com/sites /alexknapp/2015/06/12/understanding-the-thirty-meter-telescope -controversy/.

17. "History," Mauna Kea Astronomy Outreach Committee, accessed March 21, 2023, https://www.mkaoc.org/history.

18. Knapp, "Thirty Meter Telescope Controversy."

19. Kelsey Dallas, "Astronomers Are Working with Native Hawaiians to Protect a Sacred Site—and Science," *Deseret News*, August 27, 2022, https://www .deseret.com/faith/2022/8/27/23323515/mauna-kea-telescope-debate.

20. Redfield, *Space in the Tropics*; Peter Redfield, "The Half-Life of Empire in Outer Space," *Social Studies of Science* 32, no. 5/6 (2021): 791–825.

21. Fernholz and D'Auria, "Protestors Took Europe's Space Program Hostage," https://qz.com/960817/how-a-handful-of-south-american-protestors-in -french-guiana-took-arianespace-and-europes-space-program-hostage/.

22. Fernholz and D'Auria, "Protestors Took Europe's Space Program Hostage."

23. Kimberley McKinson, "Do Black Lives Matter in Outer Space?" *Sapiens*, September 3, 2020, https://www.sapiens.org/culture/space-colonization -racism/.

24. Chanda Prescod-Weinstein, *The Disordered Cosmos: A Journey into Dark Matter, Spacetime, and Dreams Deferred* (New York: Bold Type Books, 2021), 223; Sean Yeager, "A Disordered Review of Chanda Prescod-Weinstein, *The Disordered Cosmos,*" *Postmodern Culture* 31, no. 3 (2021).

25. Sheri Wells-Jenson, "The Case for Disabled Astronauts," *Scientific American*, May 30, 2018, https://blogs.scientificamerican.com/observations/the-case-for-disabled-astronauts/; Rose Eveleth, "It's Time to Rethink Who's Best Suited for Space Travel," *Wired*, January 27, 2019, https://www.wired.com/story/its-time-to-rethink-whos-best-suited-for-space-travel/.

26. Christiane Heinicke et al., "Disability in Space: Aim High," *Science* 372 (2021): 1271–1272, https://doi.org/10.1126/science.abj7353.

27. Dylan Thomas, "Do Not Go Gentle into That Good Night," *Botteghe Oscure*, 1951.

## 7. On Success and Failure

1. "Virgin Galactic Successfully Completes First Fully Crewed Spaceflight," Virgin Galactic News, July 11, 2021, https://investors.virgingalactic.com/news/news-details/2021/Virgin-Galactic-Successfully-Completes-First-Fully-Crewed-Spaceflight/default.aspx.

2. Chris Bergen, "Point-To-Point Transportation Gains Boost via NASA/Virgin Galactic SAA," NASA Spaceflight, May 5, 2020, https://www.nasaspaceflight.com/2020/05/p-t-p-transportation-boost-nasa-virgin-galactic-saa/.

3. David Valentine, Valerie Olson, and Debbora Battaglia, "Extreme: Limits and Horizons in the Once and Future Cosmos," *Anthropological Quarterly* 85, no. 4 (2012): 1007–1026.

4. Debra Craine, "Concorde, an Unexpected Success, Marks 10th Anniversary," AP News Archives, January 20, 1986, https://apnews.com/article/fa1e281d544267a8afe77afceaf3f03f (page discontinued).

5. Craine, "Concorde."

6. Doug Messier, "Virgin Galactic's Biggest Event of the Year Was a Retreat for Ticket Holders," *Parabolic Arc*, October 14, 2022, http://parabolicarc.com/2022/10/14/virgin-galactic-biggest-event-year-retreat-for-ticket-holders/ (site discontinued).

7. Virgin Hyperloop, "Smithsonian FUTURES x Virgin Hyperloop," YouTube, December 14, 2021, https://www.youtube.com/watch?v=cm2SuheeGXs&t=33s (video no longer available).

8. Bruno Latour, *Aramis, or the Love of Technology* (Cambridge: Harvard University Press, 1996).

9. Lee Vinsel and Andy Russell, *The Innovation Delusion: How Our Obsession with the New Has Disrupted the Work That Matters Most* (New York: Currency, 2020).

10. Lee Vinsel, "Prophecy and Politics, or What Are the Uses of the 'Fourth Industrial Revolution'?" Medium, March 25, 2019, https://medium.com /whats-at-stake-in-a-fourth-industrial-revolution/prophecy-and-politics-or -what-are-the-uses-of-the-fourth-industrial-revolution-30710f349ee9.

11. Vinsel, "Prophecy and Politics."

## 8. Celestial Motivations

1. "Committees," Commercial Spaceflight Federation, accessed October 27, 2022, http://www.commercialspaceflight.org/committees/.

2. See any volume of the *Handbook of STS* published through MIT Press or Nelly Oudshoorn and Trevor Pinch, eds., *How Users Matter: The Co-construction of Users and Technology* (Cambridge: MIT Press, 2005); Thomas Kuhn, *Structure of Scientific Revolutions* (Chicago: University of Chicago Press, 1962); Bruno Latour and Steve Woolgar, *Laboratory Life: The Construction of Scientific Facts* (Princeton, NJ: Princeton University Press, 1986).

3. "Mission," SpaceX, accessed November 1, 2022, https://www.spacex.com /mission/; "SpaceX's Approach to Space Sustainability and Safety," SpaceX, February 22, 2022, https://www.spacex.com/updates/; "Starship to Land NASA Astronauts on the Moon," SpaceX, April 16, 2021, https://www .spacex.com/updates/.

4. "Why Space? The Human Story, Continued," Axiom Space, accessed November 1, 2022, https://www.axiomspace.com/why-space.

5. "National Astronauts," Axiom Space, accessed November 1, 2022, https:// www.axiomspace.com/national-astronauts.

6. Rachel Kraft and Dan Huot, "NASA Names Astronauts to Next Moon Mission, First Crew Under Artemis," NASA, April 3, 2023, https://www .nasa.gov/press-release/nasa-names-astronauts-to-next-moon-mission-first -crew-under-artemis.

7. John F. Kennedy, "Address at Rice University, Houston, Texas," September 12, 1962. John F. Kennedy Presidential Library and Museum, https://www .jfklibrary.org/asset-viewer/archives/jfkpof-040-001.

8. "What We Do," Lockheed Martin, accessed November 2, 2022, https:// www.lockheedmartin.com/en-us/capabilities/space/about.html; "Protect," Lockheed Martin, accessed November 2, 2022, https://www.lockheedmartin .com/content/dam/lockheed-martin/space/documents/brochure/ Protect%20Digital.pdf; "Explore," Lockheed Martin, accessed November 2, 2022, https://www.lockheedmartin.com/content/dam/lockheed-martin /space/documents/brochure/Explore%20Digital.pdf; "Connect," Lockheed Martin, accessed November 2, 2022, https://www.lockheedmartin.com /content/dam/lockheed-martin/space/documents/brochure/Connect%20 Digital.pdf.

9. Donald Trump, "Remarks by President Trump at Kennedy Space Center," May 30, 2020, White House https://trumpwhitehouse.archives.gov /briefings-statements/remarks-president-trump-kennedy-space-center/.

10. "The Biden-Harris Administration Immediate Priorities," White House, accessed November 2, 2022, https://www.whitehouse.gov/priorities/.

11. Barack Obama, "Remarks by the President on Space Exploration in the 21st Century," Kennedy Space Center, April 15, 2010, White House, https:// obamawhitehouse.archives.gov/the-press-office/remarks-president-space -exploration-21st-century.

12. Obama, "Space Exploration in the 21st Century."

13. Obama, "Space Exploration in the 21st Century."

## 9. This Land Is Our . . .

1. Parts of this chapter have been adapted from Savannah Mandel, "Stationary Narratives: Cycles of Scientific Colonialism from Outer Space to the Tropics," *Technology and Culture* (forthcoming).

2. Megan Raby, *American Tropics: The Caribbean Roots of Biodiversity Science* (Chapel Hill: University of North Carolina Press, 2017): 57, Deborah Neill, *Networks in Tropical Medicine: Internationalism, Colonialism, and the Rise of a Medical Specialty, 1890–1930* (Stanford University Press, 2012), 2.

3. Deborah Neill, "Science and Civilizing Missions: Germans and the Transnational Community of Tropical Medicine," in *German Colonialism in a Global Age*, edited by Bradley Naranch and Geoff Eley (Durham, NC: Duke University Press, 2015).

4. Neill, "Science and Civilizing Missions."

5. Paola Villafuerte, "Decolonizing Science: What Is 'Parachute Science'?," Institute for the Future of Education, July 20, 2020, https://observatory

.tec.mx/edu-news/what-is-colonial-science; Asha de Vos, "The Problem of Colonial Science," *Scientific American*, July 1, 2020, https://www .scientificamerican.com/article/the-problem-of-colonial-science/; Michael Adas, *Dominance by Design: Technological Imperatives and America's Civilizing Mission* (Cambridge: Harvard University Press, 2006).

6. De Vos, "Problem of Colonial Science."

7. Warwick Anderson, "Introduction: Postcolonial Technoscience," *Social Studies of Science* 32, no. 5/6 (2002): 644.

8. Adas, *Dominance by Design*; Michael Adas and Hugh Glenn Cagle, "Age of Settlement and Colonisation," in *the Ashgate Research Companion to Modern Imperial Histories,* ed. John Marriot and Phillipa Levine (Milton Park, Oxfordshire: Routledge, 2012).

9. Peder Anker, "The Ecological Colonization of Space," *Environmental History* 10, no. 2 (2005): 239–268.

10. Anker, 240.

11. Anker, 240.

12. Christy Collis and Quentin Stevens, "Cold Colonies: Antarctic Spatialities at Mawson and McMurdo Stations," *Cultural Geographies* 14, no. 2 (2007): 237.

13. Hans Kohn, "Reflections on Colonialism," *Review of Politics* 18, no. 3 (1956): 259–268.

14. Jim Bleasel, "An Australian Perspective," in *Australia, Britain, and Antarctica: Papers of a Conference Held at the Australian Studies Centre* ed T. B. Millar (1986), 41;
M. Manzoni and P. Pagnini, "The Symbolic Territory of Antarctica," *Political Geography* 15, no. 5 (1996): 360.

15. Jessica O'Reilly, *The Technocratic Antarctic: An Ethnography of Scientific Expertise and Environmental Governance* (Ithaca, NY: Cornell University Press, 2017).

16. O'Reilly, *Technocratic Antarctic*.

17. Julie Klinger, "Critical Geopolitics of Outer Space," *Geopolitics* 26, no. 3 (2021): 661–665.

18. Commons in Space, November 15–17, 2023, https://2023space.iasc -commons.org/.

19. Chloe Billing, "There's a Parking Crisis in Space—and You Should Be Worried About It," *Conversation*, September 29, 2017, https:// theconversation.com/theres-a-parking-crisis-in-space-and-you-should-be -worried-about-it-83479.

20. Sir Clements R. Markham, *The Lands of Silence: A History of Arctic and Antarctic Exploration* (Cambridge: Cambridge University Press, 1921), 3.
21. Markham, 508.
22. O'Reilly, *Technocratic Antarctic*, 25.
23. Collis and Stevens, "Cold Colonies," 238.
24. O'Reilly, *Technocratic Antarctic*, 30–31.
25. O'Reilly, 33.
26. O'Reilly, 26.
27. O'Reilly, 25.
28. O'Reilly, 28.
29. O'Reilly, 27.
30. John F. Kennedy, "Special Message to the Congress on Urgent National Needs," May 25, 1961, American Presidency Project, https://www.presidency.ucsb.edu/documents/special-message-the-congress-urgent-national-needs.
31. Raby, *American Tropics*, 2.
32. Zuleyka Zevallos, "Rethinking the Narrative of Mars Colonisation," *Other Sociologist*, March 26, 2015, https://othersociologist.com/2015/03/26/rethinking-the-narrative-of-mars-colonisation/.
33. "About," Maintainers, accessed November 2, 2022, https://themaintainers.org/about/.
34. Jay Chladek, *Outposts on the Frontier: A Fifty-Year History of Space Stations* (Lincoln: University of Nebraska Press, 2017), 2.
35. Andrews and Siddiqi, *Into the Cosmos*, 1-2.
36. Andrews and Siddiqi, 1–2.
37. Andrews and Siddiqi, 1–2.
38. Andrews and Siddiqi, 2.
39. Zevallos, "Rethinking the Narrative of Mars Colonisation," https://othersociologist.com/2015/03/26/rethinking-the-narrative-of-mars-colonisation/.
40. Jeanne DiFrancesco and John Olson, "The Economics of Microgravity Research," *npj Microgravity* 1, no. 15001 (2015): 1–6.
41. "Blue Origin and Sierra Space Developing Commercial Space Stations," Blue Origin News, October 25, 2021, https://www.blueorigin.com/news/orbital-reef-commercial-space-station; "Thales Alenia Space to Provide the First Two Pressurized Modules for Axiom Space Station," Thales, press release, July 14, 2021, https://www.thalesgroup.com/en/worldwide/space/press

_release/thales-alenia-space-provide-first-two-pressurized-modules-axiom
-space.

42. "Blue Origin and Sierra Space," Thales.
43. "Blue Origin and Sierra Space," Thales.

## 10. The Exorcism of Manifest Destiny

1. US Senate, "Frequently Asked Questions About Committees," accessed November 16, 2022, https://www.senate.gov/committees/committees_faq.htm.
2. US Congress, *The Proposal to Establish a United States Space Force: Hearing, Day 1, Before Comm. on Armed Services*, 116th Cong. (2019).
3. US Senate, "Rule XXVI (d). Committee Procedure," *Standing Rules of the Senate*, revised January 24, 2013 (document 113-18), 32.
4. Congressional Research Service, *Hearings in the US Senate: A Guide for Preparation and Procedure* (March 18, 2010), 9.
5. Federal Rules of Evidence, "Rule 702. Testimony by Expert Witness," Pub. L. 93–595, §1, January 2, 1975, 88 Stat. 1937; April 17, 2000, eff. December 1, 2000; April 26, 2011, eff. December 1, 2011.
6. Latour and Woolgar, *Laboratory Life: The Construction of Scientific Facts*.
7. Steven Epstein, "The Construction of Lay Expertise: AIDS Activism and the Forging of Credibility in the Reform of Clinical Trials," *Science, Technology, & Human Values* 20, no. 4 (1995): 408–437.
8. Brian Wynne, "Misunderstood Misunderstanding: Social Identities and Public Uptake of Science." *Public Understanding of Science* 1, no. 3 (1992): 281–304.
9. US Congress, *The Proposal to Establish a United States Space Force*, 116th Cong. 15.
10. US Congress, *Discovery on the Frontiers of Space: Exploring NASA's Science Mission; Hearings, Day 1, Before Comm. on Science, Space, and Technology*, 116th Cong. (2019), 10.
11. US Congress, *Creating a Framework for Rules-based Order in Space: Hearings, Day 1, Before the Comm. on Armed Services and Committee on Foreign Affairs*, 117th Cong. (2021), 21.
12. US Congress, *A Review of NASA's Plans for the International Space Station and Future Activities in Low Earth Orbit: Hearings, Day 1, Before Comm. on Science, Space, and Technology*, 116th Cong. (2019); *A Review of the President's Fiscal Year 2022 Budget Proposal for NASA: Hearings, Day 1,*

*Before Comm. on Science, Space, and Technology*, 117th Cong. (2021); *America in Space: Future Visions, Current Issues; Hearings, Day 1, Before Comm. on Science, Space, and Technology*, 116th Cong. (2019); *America's Human Presence in Low-Earth Orbit: Hearings, Day 2, Before Comm. on Science, Space, and Technology*, 115th Cong. (2018); *Developing Core Capabilities for Deep Space Exploration: An Update on NASA's SLS, Orion, and Exploration Ground Systems; Hearings, Day 1, Before Comm. on Science, Space, and Technology*, 116th Cong. (2019); *Discovery on the Frontiers of Space*, 116th Cong.; *Event Horizon Telescope: The Black Hole Seen Round the World; Hearings, Day 1, Before Comm. on Science, Space, and Technology*, 116th Cong. (2019); *Keeping Our Sights on Mars: A Review of NASA's Deep Space Exploration Programs and Lunar Proposal; Hearings, Day 1, Before Comm. on Science, Space, and Technology*, 116th Cong. (2019); *Regulating Space: Innovation, Liberty and International Obligations; Hearings, Day 1, Before Comm. on Science, Space, and Technology*, 115th Cong. (2017); *Space Situational Awareness: Key Issues in an Evolving Landscape; Hearings, Day 2, Before Comm. on Science, Space, and Technology*, 116th Cong. (2020); *Surveying the Space Weather Landscape: Hearings, Day 2, Before Comm. on Science, Space, and Technology*, 115th Cong. (2018); *The Commercial Space Landscape: Innovation, Market, and Policy; Hearings, Day 1, Before Comm. on Science, Space, and Technology*, 116th Cong. (2019); *The Legacy of Apollo: Hearings, Day 1, Before Comm. on Science, Space, and Technology*, 116th Cong. (2019); *Moon Landings to Mars Exploration: The Role of Small Business Innovation in America's Space Program; Hearings, Day 1, Before Comm. on Small Business and Entrepreneurship*, 116th Cong. (2019); *Reopening the American Frontier: Exploring How the Outer Space Treaty Will Impact American Commerce and Settlement in Space; Hearings, Day 1, Before Comm. on Commerce, Science, and Transportation*, 115th Cong. (2017); *Near-Peer Advancements in Space and Nuclear Weapons: Hearings, Day 1, Before Comm. on Armed Services*, 117th Cong. (2021); *Military Space Operations, Policy, and Programs: Hearings, Day 1, Before Comm. on Armed Services*, 116th Cong. (2019); *The Proposal to Establish a United States Space Force*, 116th Cong.; *The Fiscal Year 2021 Air Force and Space Force Readiness Posture: Hearings, Day 2, Before Comm. on Armed Services*, 116th Cong. (2020); *Creating a Framework for Rules-Based Order in Space: Hearings, Day 1, Before Comm. on Armed Services and Comm. On Foreign Affairs*, 117th Cong. (2021).

13. *America in Space: Future Visions, Current Issues*, 116th Cong.

14. *America in Space: Future Visions, Current Issues*, 116th Cong., 51.

15. *The Legacy of Apollo*, 116th Cong., 48.

16. *Discovery on the Frontiers of Space*, 116th Cong., 11.

17. *The Proposal to Establish a United States Space Force*, 116th Cong., 47.

18. *Keeping Our Sights on Mars*, 116th Cong., 9.

19. *Military Space Operations, Policy, and Programs*, 116th Cong., 62.

20. *Military Space Operations, Policy, and Programs*, 116th Cong., 60.

21. *Creating a Framework for Rules-Based Order in Space*, 117th Cong., 13.

22. *Creating a Framework for Rules-Based Order in Space*, 117th Cong., 14.

23. *Discovery on the Frontiers of Space*, 116th Cong., 93.

24. *Military Space Operations, Policy, and Programs*, 116th Cong, 64.

25. *America in Space: Future Visions, Current Issues*, 116th Cong., 54.

## 11. They May Not Be Man

1. Dan Lockney, "NASA Technology Transfer Program," NASA, accessed December 16, 2022, https://technology.nasa.gov/; Dan Lockney, "NASA Spinoff," NASA, accessed December 16, 2022, https://spinoff.nasa.gov/.

2. "U.S. Defense Spending Compared to Other Countries," Peter G. Peterson Foundation, May 11, 2022, https://www.pgpf.org/chart-archive/0053 _defense-comparison.

3. Paul Voosen, "Trump White House Quietly Cancels NASA Research Verifying Greenhouse Gas Cuts," *Science*, May 9, 2018, https://www.science .org/content/article/trump-white-house-quietly-cancels-nasa-research -verifying-greenhouse-gas-cuts; Laura Tenenbaum, "NASA's Climate Communications Might Not Recover from the Damage of Trump's Systemic Suppression," *Time*, February 10, 2021, https://time.com/5937784 /nasa-climate-trump/; "Trump White House Axes NASA Research into Greenhouse Gas Cuts," BBC, May 10, 2018, https://www.bbc.com/news /world-us-canada-44067797.

4. "Your Guide to NASA's Budget," Planetary Society, accessed December 16, 2022, https://www.planetary.org/space-policy/nasa-budget; "NASA's FY 2022 Budget," Planetary Society, accessed December 16, 2022, https://www .planetary.org/space-policy/nasas-fy-2022-budget.

5. George Orwell, *1984* (London: Secker & Warburg, 1949).

6. Goldsmith and Rees, *The End of Astronauts: Why Robots Are the Future of Exploration* (Cambridge: Harvard University Press, 2022).

7. Goldsmith and Rees, 10.

8. Goldsmith and Rees, 8.

9. Elizabeth Wilson, *Affect and Artificial Intelligence* (Seattle: University of Washington Press, 2010); John Danaher and Neil McArthur, eds., *Robot Sex: Social and Ethical Implications* (Cambridge: MIT Press, 2017); Jennifer Robertson, *Robo Sapiens Japanicus: Robots, Gender, Family, and the Japanese Nation* (Berkeley: University of California Press, 2017); Alice Fox, "On Empathy and Alterity: How Sex Robots Encourage Us to Reconfigure Moral Status" (MA Diss., University of Twente, 2018); David Gunkel, "A Vindication of the Rights of Machines," *Philosophy of Technology* 27 (2013): 113–132; Mark Coeckelbergh, "Robot Rights? Towards a Social-Relation Justification of Moral Consideration," *Ethics of Information Technology* 12 (2010): 209–221; Vertesi, *Seeing Like a Rover*.

10. Vertesi, *Seeing Like a Rover*.

## 12. The Death of Outer Space Dreams

1. Therese Jones (@theresejones0), "I would love the future of space to be free of sexual harassment [link included]," Twitter, September 5, 2019, https://twitter.com/theresejones0/status/1169698709053030400; Luc Riesbeck (@LucRiesbeck) "The space industry has a sexual harassment problem," Twitter, April 7, 2022, https://twitter.com/lucriesbeck/status/1512081213 812404240; Chelsea Gould, "The Space Industry Has a Big, Ugly Sexual Harassment Problem," Space.com, March 30, 2022, https://www.space.com /space-industry-sexual-harassment-problem; Pandora Dewan, "Private Space Industry Has a Major Sexism Problem That's 'Too Toxic' to Fix," *Newsweek*, October 25, 2022, https://www.newsweek.com/newsweek-com -private-space-industry-major-sexism-problem-too-toxic-1754519; Joey Roulette, "Former Interns Say SpaceX Ignored Sexual Harassment," *New York Times*, December 14, 2021, https://www.nytimes.com/2021/12/14 /science/spacex-sexual-harassment.html.

2. Dewan, "Private Space Industry Has a Major Sexism Problem"; Roulette, "Former Interns Say SpaceX Ignored Sexual Harassment."

## 13. An Anthropologist's Call to Arms

1.  Mandel, "Lunar Imperialism (and How to Avoid It)."
2.  Mandel, "The Astrophile's Mite"; Mandel, "The Elysium Effect"; Savannah Mandel, "Ghosts in the Machine: On Losing Control to the Technoscape," *Platypus*, August 9, 2018, https://blog.castac.org/2018/08/ghosts/.
3.  Gutman, "A Space Anthropologist Warns: Inequality Gets Worse on Mars."
4.  Savannah Mandel, "Revisiting the Drake Equation: A Cultural Addendum," *Room Magazine* 2, no. 24 (2020); Savannah Mandel, "Under-Utilized Analogue Environments: What Can Space Scientists Learn from Arctic Communities?" *Physics Today*, December 5, 2019, https://physicstoday .scitation.org/do/10.1063/pt.6.3.20191205a/full/; Savannah Mandel, "Trial by Fire: The Legacy of Apollo 1," American Institute of Physics Publishing, press release, 2019, https://www.aip.org/news/trial-fire-legacy -apollo-1; Mandel, "The Death of Outer Space Dreams."

## 14. The Caretaker's Demand

1.  Aaron Benanav, *Automation and the Future of Work* (Brooklyn: Verso, 2020).
2.  Benanav, *Automation and the Future of Work*.
3.  Peter Frase, *Four Futures: Life After Capitalism* (London: Verso, 2016).
4.  Thomas More, *Utopia or Libellus vere aureus, nec minus salutaris quam festivus, de optimo rei publicae statu deque nova insula Utopia* (More, 1516).
5.  Nick Srnicek and Alex Williams, *Inventing the Future of Work: Postcapitalism and a World Without Work* (Brooklyn: Verso, 2016).
6.  Srnicek and Williams, *Inventing the Future of Work*.
7.  Kate Aronoff, Alyssa Battistoni, Daniel Aldana, Cohen Thea, N. Riofrancos, and Naomi Klein, *A Planet to Win: Why We Need a Green New Deal* (London: Verso, 2019); Noam Chomsky, Robert Pollin, and Chronis Polychroniou, *Climate Crisis and the Global Green New Deal: The Political Economy of Saving the Planet* (London: Verso, 2020); Naomi Klein, *On Fire: The (Burning) Case for a Green New Deal* (New York: Simon & Schuster, 2020).
8.  Marianna Brady, "Coronavirus: How the Pandemic Is Relaxing US Drinking Laws," BBC News, May 15, 2020, https://www.bbc.com/news/world-us -canada-52656756.
9.  Kim Stanley Robinson, *Aurora* (London: Orbit, 2015); Orson Scott Card, *Ender's Game* (New York: Tor Books, 1985); Joe Haldeman, *The Forever War* (New York: St. Martin's Press, 1974).

10. Kim Stanley Robinson, *The Ministry for the Future* (London: Orbit, 2020).

## Epilogue: Earthward Auguries and Activisms

1.  Cory-Alice André-Johnson, "What Does Anthropology Sound Like:
    Activism," *Anthropod*, January 20, 2020, https://culanth.org/fieldsights/what
    -does-anthropology-sound-like-activism; Charles R. Hale, "Activist Research
    v. Cultural Critique: Indigenous Land Rights and the Contradictions of
    Politically Engaged Anthropology," *Cultural Anthropology* 21, no. 1 (2008):
    96–101; Anna J. Willow and Kelly A. Yotebieng, eds. *Anthropology and
    Activism: New Contexts, New Conversations* (New York: Routledge, 2020).

# BIBLIOGRAPHY

Abrahamian, Atossa Araxia. "How the Asteroid-Mining Bubble Burst."
*MIT Review*, June 26, 2019. https://www.technologyreview.
com/2019/06/26/134510/asteroid-mining-bubble-burst-history/.

Adas, Michael. *Dominance by Design: Technological Imperatives and America's Civilizing Mission.* Cambridge: Harvard University Press, 2006.

Adas, Michael, and Hugh Glenn Cagle. "Age of Settlement and Colonisation."
In *The Ashgate Research Companion to Modern Imperial Histories*, edited by John Marriot and Phillipa Levine. Milton Park, Oxfordshire: Routledge, 2012.

Aiken, Jo, and Angela Ramer. "From the Space Station to the Sofa: Scales of Isolation at Work." *Ethnographic Praxis in Industry Conference Proceedings*, October 2020. https://doi.org/10.1111/epic.12044.

Anderson, Warwick. "Introduction: Postcolonial Technoscience." *Social Studies of Science* 32, no. 5/6 (2002): 644.

André-Johnson, Cory-Alice. "What Does Anthropology Sound Like: Activism."
*Anthropod*, January 20, 2020. https://culanth.org/fieldsights/what-does
-anthropology-sound-like-activism.

Andrews, James, and Asif Siddiqi, eds. *Into the Cosmos: Space Exploration and Soviet Culture.* Pittsburgh: University of Pittsburgh Press, 2011.

Anker, Peder. "The Ecological Colonization of Space." *Environmental History* 10, no. 2 (2005): 239–268.

Aristophanes. *The Frogs.* In *The Complete Greek Drama: All the Extant Tragedies of Aeschylus, Sophocles and Euripides, and the Comedies of Aristophanes and Menander, in a Variety of Translations*, edited by Whitney J. Oates and Eugene O'Neill. New York: Random House, 1938.

Aronoff, Kate, Alyssa Battistoni, Daniel Aldana, Cohen Thea, N. Riofrancos, and Naomi Klein. *A Planet to Win: Why We Need a Green New Deal*. London: Verso, 2019.

Associated Press. "Former New Mexico Spaceport CFO Alleges Fraud, Retaliation." *US News & World Report*, January 5, 2022. https://www.usnews.com/news/best-states/new-mexico/articles/2022-01-05/former-new-mexico-spaceport-cfo-alleges-fraud-retaliation.

Axiom Space. "National Astronauts." Axiom Space. Accessed November 1, 2022. https://www.axiomspace.com/national-astronauts.

———. "Why Space? The Human Story, Continued." Accessed November 1, 2022. https://www.axiomspace.com/why-space.

Barnd, Natchee Blu. *Native Space: Geographic Strategies to Unsettle Settler Colonialism*. Corvallis: Oregon State University Press, 2017.

Battaglia, Debbora. "Arresting Hospitality: The Case of the 'Handshake in Space.'" *Journal of the Royal Anthropological Institute* 18 (2012): S76–S89.

———. "Diary of a Space Zucchini: Ventriloquizing the Future in Outer Space." *Platypus*, July 14, 2014. https://blog.castac.org/2014/07/diary-of-a-space-zucchini-ventriloquizing-the-future-in-outer-space/.

———. *E.T. Culture: Anthropology in Outerspaces*. Durham, NC: Duke University Press, 2006.

BBC. "Trump White House Axes NASA Research into Greenhouse Gas Cuts." May 10, 2018. https://www.bbc.com/news/world-us-canada-44067797.

Becenti, Francis. "Going into Space." 1971.

Benanav, Aaron. *Automation and the Future of Work*. Brooklyn: Verso, 2020.

Bergen, Chris. "Point-to-Point Transportation Gains Boost via NASA/Virgin Galactic SAA." NASA Spaceflight, May 5, 2020. https://www.nasaspaceflight.com/2020/05/p-t-p-transportation-boost-nasa-virgin-galactic-saa/.

Bieri, Kate. "How Much Has Spaceport America Cost Taxpayers?" KVIA.com, July 9, 2021. https://kvia.com/news/new-mexico/2021/07/09/how-much-has-spaceport-building-cost-dona-ana-county-taxpayers/.

Billing, Chloe. "There's a Parking Crisis in Space—and You Should Be Worried About It." *Conversation*, September 29, 2017. https://theconversation.com/theres-a-parking-crisis-in-space-and-you-should-be-worried-about-it-83479.

Bleasel, Jim. "An Australian Perspective." In *Australia, Britain, and Antarctica: Papers of a Conference Held at the Australian Studies Centre*, edited by T. B. Millar. 1986.

Blue Origin News. "Blue Origin and Sierra Space Developing Commercial Space Stations." October 25, 2021. https://www.blueorigin.com/news/orbital-reef -commercial-space-station.

Booker, Simeon. "Moon Probe Laudable—but Blacks Need Help." *Jet*, August 1, 1969, 10.

Bourgois, Philippe, and Jeffrey Schonberg. *Righteous Dopefiend*. Berkeley: University of California Press, 2009.

Brady, Marianna. "Coronavirus: How the Pandemic Is Relaxing US Drinking Laws." BBC News, May 15, 2020. https://www.bbc.com/news/world-us -canada-52656756.

Buchli, Victor. 'Extra-terrestrial Methods: Towards an Ethnography of the ISS." In *Lineages and Advancements in Material Culture Studies: Perspective from UCL Anthropology*, edited by T. Carroll et al. London: Bloomsbury, forthcoming.

———. "Low Earth Orbit: A Speculative Ethnographer's Guide." In *Anti-Atlas: Towards a Critical Area Studies*, edited by W. Bracewell et al. London: UCL Press, forthcoming.

Burrington, Ingrid. "New Mexico's Sad Bet on Space Exploration." *Atlantic*, March 2, 2018. https://www.theatlantic.com/technology/archive/2018/03 /new-mexicos-sad-bet-on-space-exploration/554243/.

Callahan, David, and Fred Greenstein. "The Reluctant Racer: Eisenhower and U.S. Space Policy." In *Spaceflight and the Myth of Presidential Leadership.*, edited by Roger Lanius and Howard McCurdy. University of Illinois Press, 1997.

Canneti, Elias. *Crowds and Power*. London: Victor Gollancz, 1962.

Card, Orson Scott. *Ender's Game*. New York: Tor Books, 1985.

Carroll, Timothy, and Aaron Parkhurst, "Being, Being Human, Becoming Beyond Human." In *Lineages and Advancements in Material Culture Studies*. London: Routledge, 2020.

Carruthers, Charlene. "Hearing Assata Shakur's Call." *Women's Studies Quarterly* 46, no. 3/4 (2018): 222–225.

Carter, Jamie. "Are We Finally on the Cusp of Commercial Asteroid Mining?" *Sky & Telescope*, February 27, 2023. https://skyandtelescope.org/astronomy -news/are-we-finally-on-the-cusp-of-commercial-asteroid-mining/.

Chaturvedi, Sanjay. *The Polar Regions: A Political Geography*. New York: Wiley, 1996.

Cheney, Matthew. "The Good, The Bad, and the Interdisciplinary." *Finite Eyes*, October 3, 2021. https://finiteeyes.net/interdisciplinary/the-good-the-bad -and-the-interdisciplinary/.

Chladek, Jay. *Outposts on the Frontier: A Fifty-Year History of Space Stations.*
    Lincoln: University of Nebraska Press, 2017.

Chomsky, Noam, Robert Pollin, and Chronis Polychroniou. *Climate Crisis and
    the Global Green New Deal: The Political Economy of Saving the Planet.*
    London: Verso, 2020.

Coeckelbergh, Mark. "Robot Rights? Towards a Social-Relation Justification of
    Moral Consideration." *Ethics of Information Technology* 12 (2010): 209–221.

Collis, Christy, and Quentin Stevens. "Cold Colonies: Antarctic Spatialities at
    Mawson and McMurdo Stations." *Cultural Geographies* 14, no. 2 (2007): 237.

Commercial Spaceflight Federation. "Committees." Accessed October 27, 2022.
    http://www.commercialspaceflight.org/committees/.

Commons in Space, November 15–17, 2023. https://2023space.iasc-commons
    .org/.

Congressional Research Service. *Hearings in the U.S. Senate: A Guide for
    Preparation and Procedure.* March 18, 2010.

Craine, Debra. "Concorde, An Unexpected Success, Marks 10th Anniversary."
    AP News Archives, January 20, 1986. https://apnews.com/article
    /fa1e281d544267a8afe77afceaf3f03f (page discontinued).

Creighton, Jolene. "History's Future in Space Depends on Asteroid Mining."
    *Futurism*, June 23, 2016. https://futurism.com/humanitys-future-in-space
    -depends-on-asteroid-mining.

Cruikshank, Julie. "Are Glaciers 'Good to Think With'? Recognising Indigenous
    Environmental Knowledge." *Anthropological Forum* 22, no. 3 (2012):
    239–250.

Dallas, J. A., S. Raval, J. P. Alvarez Gaitan, S. Saydam, and A. G Dempster. "Mining
    Beyond Earth for Sustainable Development: Will Humanity Benefit from
    Resource Extraction in Outer Space?" *Acta Astronautica* 167 (2020): 181–188.

Dallas, Kelsey. "Astronomers Are Working with Native Hawaiians to Protect a
    Sacred Site—and Science." *Deseret News*, August 27, 2022. https://www
    .deseret.com/faith/2022/8/27/23323515/mauna-kea-telescope-debate.

Danaher, John, and Neil McArthur eds. *Robot Sex: Social and Ethical
    Implications.* Cambridge: MIT Press, 2017.

De Vos, Asha. "The Problem of Colonial Science." *Scientific American*, July 1,
    2020. https://www.scientificamerican.com/article/the-problem-of-colonial
    -science/.

Dewan, Pandora. "Private Space Industry Has a Major Sexism Problem That's
    'Too Toxic' to Fix." *Newsweek*, October 25, 2022. https://www.newsweek

.com/newsweek-com-private-space-industry-major-sexism-problem-too
-toxic-1754519.

Diaz, Natalie. *Postcolonial Love Poems*. Minneapolis: Graywolf Press, 2020.

DiFrancesco, Jeanne, and John Olson. "The Economics of Microgravity
Research." *npj Microgravity* 1, 15001 (2015).

Dorminey, Bruce. "Does Commercial Asteroid Mining Still Have a
Future?' *Forbes*, August 31, 2021. https://www.forbes.com/sites/
brucedorminey/2021/08/31/does-commercial-asteroid-mining-still-have-a
-future/.

Dunietz, Jesse. "Floating Treasure: Space Law Needs to Catch Up with Asteroid
Mining." *Scientific American*, August 28, 2017. https://www.scientificamerican
.com/article/floating-treasure-space-law-needs-to-catch-up-with-asteroid
-mining/.

Eisenhower, Dwight D. "Address Before the 15th General Assembly of the United
Nations," September 22, 1960. US Department of State Archived Content.
https://2009-2017.state.gov/p/io/potusunga/207330.htm.

Emerson, Robert, Rachel Fretz, and Linda Shaw. *Writing Ethnographic Fieldnotes*.
Chicago: University of Chicago Press, 1995.

Epstein, Steven. "The Construction of Lay Expertise: AIDS Activism and the
Forging of Credibility in the Reform of Clinical Trials." *Science, Technology,
& Human Values* 20, no. 4 (1995): 408–437.

Ethno-ISS. "About." Accessed September 14, 2022. https://ethnoiss.space/.

Etzioni, Amitai. *The Moon-Doggle: Domestic and International Implications of the
Space Race*. New York: Doubleday, 1964.

European Commission. "ETHNO-ISS: An Ethnography of an Extra-terrestrial
Society: The International Space Station." September 4, 2019. https://cordis.
europa.eu/project/id/833135.

Eveleth, Rose. "It's Time to Rethink Who's Best Suited for Space Travel." *Wired*,
January 27, 2019. https://www.wired.com/story/its-time-to-rethink-whos
-best-suited-for-space-travel/.

Federal Rules of Evidence. "Rule 702. Testimony by Expert Witness." Pub. L.
93–595, §1, January 2, 1975, 88 Stat. 1937; April 17, 2000, eff. December 1,
2000; April 26, 2011, eff. December 1, 2011.

Fernholz, Tim, and Peter D'Auria. "How a Handful of South American Protestors
Took Europe's Space Program Hostage." *Quartz*, April 17, 2017. https://
qz.com/960817/how-a-handful-of-south-american-protestors-in-french
-guiana-took-arianespace-and-europes-space-program-hostage/.

Foust, Jeff. "Virgin Galactic and the Future of Commercial Spaceflight." Space.
    com, May 23, 2005. https://www.space.com/1110-virgin-galactic-future
    -commercial-spaceflight.html.

Fox, Alice. "On Empathy and Alterity: How Sex Robots Encourage Us to
    Reconfigure Moral Status." MA Diss., University of Twente, 2018.

Frase, Peter. *Four Futures: Life After Capitalism*. London: Verso, 2016.

Geertz, Clifford. "Deep Play: Notes on the Balinese Cockfight." *Daedalus* 101, no.
    1 (1972): 1–37.

Gilbert, Alex. "Mining in Space Is Coming." *Milken Institute Review*, April 26,
    2021. https://www.milkenreview.org/articles/mining-in-space-is-coming.

Goffman, Alice. *On the Run: Fugitive Life in an American City*. Chicago:
    University of Chicago Press, 2014.

Goguichvili, Sophie, Alan Linenberger, Amber Gillette, et al. "The Global Legal
    Landscape of Space: Who Writes the Rules on the Final Frontier?' Wilson
    Center, October 1, 2021. https://www.wilsoncenter.org/article/global-legal
    -landscape-space-who-writes-rules-final-frontier.

Goldsmith, Donald, and Martin Rees. *The End of Astronauts: Why Robots Are the
    Future of Exploration*. Cambridge: Harvard University Press, 2022.

Gould, Chelsea. "Report Finds That Former Spaceport America Director
    Violated State Laws (Report)." Space.com, December 4, 2020. https://www
    .space.com/spaceport-america-former-director-investigation.

———. "The Space Industry Has a Big, Ugly Sexual Harassment Problem." Space.
    com, March 30, 2022. https://www.space.com/space-industry-sexual
    -harassment-problem.

Greshko, Michael. "China's 'Jade Rabbit' Moon Rover Declared Dead." *National
    Geographic*, August 4, 2016. https://www.nationalgeographic.com/science
    /article/china-moon-rover-jade-rabbit-dead-yutu-space-science.

Grimason, Maggie. "Spaceport America: New Mexico's Protracted Hope for a
    Revived Economy." *Wire Science*, May 8, 2018. https://science.thewire.in
    /aerospace/spaceport-america-new-mexicos-protracted-hope-for-a-revived
    -economy/.

Gunkel, David. "A Vindication of the Rights of Machines." *Philosophy of
    Technology* 27 (2013): 113–132.

Gurney, James. *Dinotopia, a Land Apart from Time: 20th Anniversary Edition*.
    Mineola: Calla Editions, 2011.

Gutman, Leslie. "A Space Anthropologist Warns: Inequality Gets Worse on
    Mars." *Ozy*, October 24, 2019. https://www.ozy.com/the-new-and-the-next

/a-space-anthropologist-warns-inequality-only-gets-worse-on-mars/221963 / (page discontinued).

Haldeman, Joe. *The Forever War*. New York: St. Martin's Press, 1974.

Hale, Charles R. "Activist Research v. Cultural Critique: Indigenous Land Rights and the Contradictions of Politically Engaged Anthropology." *Cultural Anthropology* 21, no. 1 (2008): 96–101.

Heinicke, Christiane et al. "Disability in Space: Aim High." *Science* 372 (2021): 1271–1272. https://doi.org/10.1126/science.abj7353.

Heise, Jack. "Space, the Final Frontier of Enterprise: Incentivizing Asteroid Mining Under a Revised International Framework." *Michigan Journal of International Law* 40, no. 1 (2018): 189.

Helmreich, Stefan. *Alien Ocean: Anthropological Voyages in Microbial Seas*. Berkeley: University of California Press, 2009.

Hesiod. *Works and Days*. Translated by H. G. Evelyn-White. In Loeb Classical Library, vol. 57. London: William Heinemann, 1914.

Hindman, Matthew Dean. "Interdisciplinarity's Shared Governance Problem." American Association of University Professors, accessed December 29, 2022. https://www.aaup.org/article/interdisciplinarity%E2%80%99s-shared -governance-problem.

Homer. *The Odyssey*. Translated by A. T. Murray. Cambridge: Harvard University Press, 1919.

Hu, Jane C. "First Man Offers a Helpful Reminder That a Lot of Americans Didn't Want to Go to the Moon." *Quartz*, October 25, 2018. https:// qz.com/1432303/first-man-shows-that-many-americans-opposed-nasas -moon-mission/.

Ishola, Feyisola Ruth, Oluwabusola Fadipe, and Olaoluwa Colin Taiwo. "Legal Enforceability of International Space Laws: An Appraisal of 1967 Outer Space Treaty." *New Space* 9, no. 1 (2021): 33–37.

ISS National Laboratory. "An Unparalleled Environment for Research and Development." Accessed November 4, 2021. https://www.issnationallab.org /research-on-the-iss/iss-research-advantages.

Jacobs, Jerry A. "Interdisciplinary Hype." *Chronicle of Higher Education*, November 22, 2009. https://www.chronicle.com/article/interdisciplinary -hype/.

Jeevendrampillai, David, and Aaron Parkhurst. "Making a Martian Home: Finding Humans on Mars Through Utopian Architecture." *Home Cultures* (2021): 1751–7427.

———. "Towards an Anthropology of Gravity: Emotion and Embodiment in Microgravity Environments." *Emotion, Space & Society* 35 (2020).

Johnson, Thomas A. "Blacks and Apollo: Most Couldn̄t Have Cared Less." *New York Times*, July 27, 1969.

Jones, Therese (@theresejones0). "I would love the future of space to be free of sexual harassment [link included]." Twitter, September 5, 2019. https://twitter.com/theresejones0/status/1169698709053030400.

Kahanamoku, Sara, Rosie 'Anolani Alegado, Aurora Kagawa-Viviani, Katie Leimomi Kamelamela, Brittany Kamai, Lucianne M. Walkowicz, Chanda Prescod-Weinstein, Mithi Alexa de los Reyes, and Hilding Neilson. "A Native Hawaiian–Led Summary of the Current Impact of Constructing the Thirty Meter Telescope on Maunakea." White paper submitted to the National Academy of Sciences Decadal Survey on Astronomy and Astrophysics Panel on the State of the Profession and Societal Impacts, January 3, 2020.

Kennedy, John. F. "Address at Rice University, Houston, Texas," September 12, 1962. John F. Kennedy Presidential Library and Museum. https://www.jfklibrary.org/asset-viewer/archives/jfkpof-040-001.

———. "Special Message to the Congress on Urgent National Needs," May 25, 1961. American Presidency Project. https://www.presidency.ucsb.edu/documents/special-message-the-congress-urgent-national-needs.

Klein, Naomi. *On Fire: The (Burning) Case for a Green New Deal*. New York: Simon & Schuster, 2020.

Klinger, Julie. "Critical Geopolitics of Outer Space." *Geopolitics* 26, no. 3 (2021): 661–665.

Knapp, Alex. "Understanding the Thirty Meter Telescope Controversy." *Forbes*, June 12, 2015. https://www.forbes.com/sites/alexknapp/2015/06/12/understanding-the-thirty-meter-telescope-controversy/.

Kohn, Hans. "Reflections on Colonialism." *Review of Politics* 18, no. 3 (1956): 259–268.

Kraft, Rachel, and Dan Huot. "NASA Names Astronauts to Next Moon Mission, First Crew Under Artemis." NASA, April 3, 2023. https://www.nasa.gov/press-release/nasa-names-astronauts-to-next-moon-mission-first-crew-under-artemis.

Kuhn, Thomas. *Structure of Scientific Revolutions*. Chicago: University of Chicago Press, 1962.

Latour, Bruno. *Aramis, or the Love of Technology*. Cambridge: Harvard University Press, 1996.

Latour, Bruno, and Steve Woolgar. *Laboratory Life: The Construction of Scientific Facts*. Princeton, NJ: Princeton University Press, 1986.

Le Guin, Ursula K. 'The Ones Who Walk Away from Omelas.' *New Dimensions* 3 (1973).

Lockheed Martin. "Connect." Accessed November 2, 2022. https://www .lockheedmartin.com/content/dam/lockheed-martin/space/documents /brochure/Connect%20Digital.pdf.

———. "Explore." Accessed November 2, 2022. https://www.lockheedmartin .com/content/dam/lockheed-martin/space/documents/brochure /Explore%20Digital.pdf.

———. "Protect." Accessed November 2, 2022. https://www.lockheedmartin .com/content/dam/lockheed-martin/space/documents/brochure /Protect%20Digital.pdf.

———. "What We Do." Accessed November 2, 2022. https://www .lockheedmartin.com/en-us/capabilities/space/about.html.

Lockney, Dan. "NASA Spinoff." NASA. Accessed December 16, 2022. https:// spinoff.nasa.gov/.

———. "NASA Technology Transfer Program." NASA. Accessed December 16, 2022. https://technology.nasa.gov/.

Maher, Neil M. *Apollo During the Age of Aquarius*. Cambridge: Harvard University Press, 2019.

Maintainers. "About." Accessed November 2, 2022. https://themaintainers.org /about/.

Mandel, Savannah. "The Astrophile's Mite: Risk and Sacrifice in the Commercial Space Industry." MSc diss., University College London, 2018.

———. "The Death of Outer Space Dreams: Hard Decisions and a War of Utopian Demands." *Geek Anthropologist*, April 22, 2021. https:// thegeekanthropologist.com/2021/04/22/the-death-of-outer-space-dreams -hard-decisions-and-a-war-of-utopian-demands/.

———. "The Elysium Effect: Space Law and Commercial Space Disparities." *Geek Anthropologist*, July 10, 2018. https://thegeekanthropologist.com/2018/07/10 /the-elysium-effect-thoughts-on-that-future-we-dont-want-to-happen/.

———. "Ghosts in the Machine: On Losing Control to the Technoscape." *Platypus*, August 9, 2018. https://blog.castac.org/2018/08/ghosts/.

———. "Lunar Imperialism (and How to Avoid It)." *Association for Political and Legal Anthropology*, July 12, 2019. https://www.anthropology-news.org /articles/lunar-imperialism-and-how-to-avoid-it/.

———. "On Refrigerators and Rage: Secrecy and Pacification in the Florida Restaurant Industry." *Journal of Contemporary Ethnography* 52, no. 1 (2021).

———. "Revisiting the Drake Equation: A Cultural Addendum." *Room Magazine* 2, no. 24 (2020).

———. "Stationary Narratives: Cycles of Scientific Colonialism from Outer Space to the Tropics." *Technology and Culture*, forthcoming.

———. "Trial by Fire: The Legacy of Apollo 1." American Institute of Physics Publishing, press release, 2019. https://www.aip.org/news/trial-fire-legacy -apollo-1.

———. "Under-Utilized Analogue Environments: What Can Space Scientists Learn from Arctic Communities?" *Physics Today*, December 5, 2019. https:// physicstoday.scitation.org/do/10.1063/pt.6.3.20191205a/full/.

Manzoni, M., and P. Pagnini. "The Symbolic Territory of Antarctica." *Political Geography* 15, no. 5 (1996): 360.

Mar, Tracey Banivanua, and Penelope Edmonds eds. *Making Settler Colonial Space: Perspectives on Race, Place and Identity.* London: Palgrave Macmillan, 2010.

Markham, Sir Clements R. *The Lands of Silence: A History of Arctic and Antarctic Exploration.* Cambridge: Cambridge University Press, 1921.

Martin, John Levi. *Thinking Through Methods: A Social Science Primer.* Chicago: University of Chicago Press, 2017.

Maruyama, Magoroh, and Arthur Harkins. *Cultures Beyond Earth: The Role of Anthropology in Outer Space.* New York: Vintage House, 1975.

Mauna Kea Astronomy Outreach Committee. "History." Accessed March 21, 2023. https://www.mkaoc.org/history.

McDougall, Walter. . . . *The Heavens and the Earth: A Political History of the Space Age.* Baltimore, MD: John Hopkins University Press, 1985.

McGowan, James. *Goodnight Oppy.* New York: Astra Young Readers, 2021.

McKinson, Kimberley. "Do Black Lives Matter in Outer Space?" *Sapiens,* September 3, 2020. https://www.sapiens.org/culture/space-colonization -racism/.

Messeri, Lisa. "Extra-Terra Incognita: Martian Maps in the Digital Age." *Social Studies of Science* 47, no. 1 (2016): 75–94.

Messier, Doug. "A Brief History of Spaceport America." *Parabolic Arc*, August 19, 2019. http://www.parabolicarc.com/2019/08/19/a-brief-history-of -spaceport-america/.

———. "NM's Latest Plan to Make Money from Spaceport America: Stop Treating SpaceShipTwo Passengers as Freight." *Parabolic Arc*, January 27, 2022. http://

www.parabolicarc.com/2022/01/27/nm-plans-money-spaceport-america
-involves-longer-treating-spaceshiptwo-passengers-freight/.

———. "Virgin Galactic's Biggest Event of the Year Was a Retreat for Ticket
Holders." *Parabolic Arc*, October 14, 2022. http://parabolicarc.com/2022
/10/14/virgin-galactic-biggest-event-year-retreat-for-ticket-holders/.

Molina, David Miguel, and P. J. Blount. "Bringing the Moon to Mankind: The
Civil Rights Narrative and the Space Age." In *NASA and the Long Civil Rights
Movement* (Gainesville: University Press of Florida, 2019).

More, Thomas. *Utopia or Libellus vere aureus, nec minus salutaris quam festivus,
de optimo rei publicae statu deque nova insula Utopia*. More, 1516.

Neill, Deborah. *Networks in Tropical Medicine: Internationalism, Colonialism,
and the Rise of a Medical Specialty, 1890–1930*. Redwood City, CA: Stanford
University Press, 2012.

———. "Science and Civilizing Missions: Germans and the Transnational
Community of Tropical Medicine." In *German Colonialism in a Global Age*,
edited by Bradley Naranch and Geoff Eley. Durham, NC: Duke University
Press, 2015.

Niiler, Eric. "Why Civil Rights Activists Protested the Moon Landing." History,
July 11, 2019. https://www.history.com/news/apollo-11-moon-landing
-launch-protests.

Obama, Barack. "Remarks by the President on Space Exploration in the 21st
Century," April 15, 2010. White House. https://obamawhitehouse.archives.
gov/the-press-office/remarks-president-space-exploration-21st-century.

Olson, Valerie. "American Extreme: An Ethnography of Astronautical Visions
and Ecologies." PhD diss., Rice University, 2010.

———. *Into the Extreme: U.S. Environmental Systems and Politics Beyond Earth*.
Minneapolis: University of Minnesota Press, 2018.

Oman-Reagan, Michael. "Anthropologists in Outer Space." *Wanderers*, February
8, 2016. https://www.sapiens.org/culture/anthropologists-in-outer-space/.

O'Reilly, Jessica. *The Technocratic Antarctic: An Ethnography of Scientific Expertise
and Environmental Governance*. Ithaca, NY: Cornell University Press, 2017.

Orwell, George. *1984*. London: Secker & Warburg, 1949.

Oudshoorn, Nelly, and Trevor Pinch, eds., *How Users Matter: The Co-
construction of Users and Technology*. Cambridge: MIT Press, 2005.

Parkhurst, Aaron, and David Jeevendrampillai. "Towards an Anthropology
of Gravity: Emotion and Embodiment in Microgravity Environments."
*Emotion, Space and Society* 35 (2020).

Peter G. Peterson Foundation. "U.S. Defense Spending Compared to Other Countries." May 11, 2022. https://www.pgpf.org/chart-archive/0053 _defense-comparison.

Peterson, Valerie V. "Against Interdisciplinarity." *Women and Language* 31, no. 2 (2008): 42–50.

Planetary Society. "NASA's FY 2022 Budget." Accessed December 16, 2022. https://www.planetary.org/space-policy/nasas-fy-2022-budget.

———. "Your Guide to NASA's Budget." Accessed December 16, 2022. https://www.planetary.org/space-policy/nasa-budget.

Plutarch. *The Parallel Lives*. Translated by Bernadotte Perrin. In Loeb Classical Library, vol. 8. Cambridge: Harvard University Press, 1919.

Prescod-Weinstein, Chanda. *The Disordered Cosmos: A Journey into Dark Matter, Spacetime, and Dreams Deferred*. New York: Bold Type Books, 2021.

Raby, Megan. *American Tropics: The Caribbean Roots of Biodiversity Science*. Chapel Hill: University of North Carolina Press, 2017.

Rao, Sonia. "Why 'First Man' Prominently Features Gil Scott-Heron's Spoken-Word Poem 'Whitey on the Moon.'" *Washington Post*, October 13, 2018. https://www.washingtonpost.com/arts-entertainment/2018/10/13/why-first -man-prominently-features-gil-scott-herons-spoken-word-poem-whitey -moon/.

Redfield, Peter. "The Half-Life of Empire in Outer Space." *Social Studies of Science* 32, no. 5/6 (2021): 791–825.

———. *Space in the Tropics: From Convicts to Rockets in French Guiana*. Berkeley: University of California Press, 2000.

Reuter, Dominik. "600 People Have Reserved $250,000 Tickets to Fly to Space with Virgin Galactic, Including Celebrities like Tom Hanks, Leonardo DiCaprio, Justin Bieber, and Lady Gaga." *Business Insider India*, July 10, 2021. https://www.businessinsider.in/tech/news/600-people-have-reserved -250000-tickets-to-fly-to-space-with-virgin-galactic-including-celebrities -like-tom-hanks-leonardo-dicaprio-justin-bieber-and-lady-gaga /articleshow/84281007.cms.

Riesbeck, Luc (@LucRiesbeck). "The space industry has a sexual harassment problem." Twitter, April 7, 2022. https://twitter.com/lucriesbeck /status/1512081213812404240.

Robertson, Jennifer. *Robo Sapiens Japanicus: Robots, Gender, Family, and the Japanese Nation*. Berkley: University of California Press, 2017.

Robinson, Kim Stanley. *Aurora*. London: Orbit, 2015.

———. *The Ministry for the Future*. London: Orbit, 2020.

Rogers, Adam. "Space Tourism Isn't Worth Dying For." *Wired*, October 31, 2014. https://www.wired.com/2014/10/virgin-galactic-boondoggle/.

Rosenblum, Andrew. "How to Mine an Asteroid." *Popular Science*, December 17, 2011. https://www.popsci.com/science/article/2011-11/how-mine-asteroid/.

Roulette, Joey. "Former Interns Say SpaceX Ignored Sexual Harassment." *New York Times*, December 14, 2021. https://www.nytimes.com/2021/12/14/science/spacex-sexual-harassment.html.

Schmidle, Nicholas. *Test Gods: Virgin Galactic and the Making of the Modern Astronaut*. New York: Henry Holt and Co., 2021.

Scott-Heron, Gil. "Whitey on the Moon." Track 9 on *Small Talk at 125th and Lenox*. Flying Dutchman Records, 1970.

Shakur, Assata. *Assata: An Autobiography*. Chicago: Chicago Review Press, 1999.

Siddiqi, Asif. *The Red Rockets' Glare: Spaceflight and the Soviet Imagination, 1857–1957*. Cambridge: Cambridge University Press, 2010.

Smiles, Deondre. "The Settler Logics of (Outer) Space." *Society & Space*, October 26, 2020. https://www.societyandspace.org/articles/the-settler-logics-of-outer-space.

SpaceX. "Mission." Accessed November 1, 2022. https://www.spacex.com/mission/.

———. "SpaceX's Approach to Space Sustainability and Safety." February 22, 2022. https://www.spacex.com/updates/.

———. "Starship to Land NASA Astronauts on the Moon." April 16, 2021. https://www.spacex.com/updates/.

Srnicek, Nick, and Alex Williams. *Inventing the Future of Work: Postcapitalism and a World Without Work*. Brooklyn: Verso, 2016.

Squire, Rachael, Oli Mould, and Peter Adey. "The Final Frontier? The Enclosure of a Commons of Outer Space." *Society and Space* 39, no. 5 (2021). https://www.societyandspace.org/forums/the-final-frontier-the-enclosure-of-a-commons-of-outer-space.

Tenenbaum, Laura. "NASA's Climate Communications Might Not Recover from the Damage of Trump's Systemic Suppression." *Time*, February 10, 2021. https://time.com/5937784/nasa-climate-trump/.

Thales. "Thales Alena Space to Provide the First Two Pressurized Modules for Axiom Space Station." Press release, July 14, 2021. https://www.thalesgroup.com/en/worldwide/space/press_release/thales-alenia-space-provide-first-two-pressurized-modules-axiom-space.

Thomas, Dylan. "Do Not Go Gentle into That Good Night." *Botteghe Oscure*, 1951.

*Times*. "Americans Favour End to Moon Race." April 9, 1963.

Trump, Donald. "Remarks by President Trump at Kennedy Space Center," May 30, 2020. White House. https://trumpwhitehouse.archives.gov/briefings -statements/remarks-president-trump-kennedy-space-center/.

*Truth or Consequences*. Directed by Hannah Jayanti. Sentient.Art.Film, 2020.

United Nations. "Agreement Governing the Activities of States on the Moon and Other Celestial Bodies." Resolution 34/68. Opened for signature December 18, 1979. https://www.unoosa.org/oosa/en/ourwork/spacelaw/treaties.html.

———. "Treaty on Principles Governing the Activities of States in the Exploration and Use of Outer Space, Including the Moon and Other Celestial Bodies." Resolution 2222 (XXI). Opened for signature January 27, 1967. https://www.unoosa.org/oosa/en/ourwork/spacelaw/treaties.html.

United Press International. "NASA Chief Briefs Abernathy After Protest at Cape." July 16, 1969. https://www.upi.com/Archives/1969/07/16/NASA-chief-briefs -Abernathy-after-protest-at-Cape/7371558396299/.

US Congress. *America in Space: Future Visions, Current Issues; Hearings, Day 1, Before Comm. on Science, Space, and Technology*. 116th Cong. (2019).

———. *America's Human Presence in Low-Earth Orbit: Hearings, Day 2, Before Comm. on Science, Space, and Technology*. 115th Cong. (2018).

———. *The Commercial Space Landscape: Innovation, Market, and Policy; Hearings, Day 1, Before Comm. on Science, Space, and Technology*. 116th Cong. (2019).

———. Commercial Space Launch Act. H.R. 3942, 98th Cong. (1983–1984). https://www.congress.gov/bill/98th-congress/house-bill/3942/text.

———. Commercial Space Launch Amendments Act. H.R. 5382, 108th Cong. (2003–2004). https://www.congress.gov/bill/108th-congress/house -bill/5382/text.

———. Commercial Space Launch Competitiveness Act. H.R .2262, 114th Cong. (2015–2016). https://www.congress.gov/bill/114th-congress/house-bill/2262 /text.

———. *Creating a Framework for Rules-Based Order in Space: Hearings, Day 1, Before the Comm. on Armed Services and Committee on Foreign Affairs*. 117th Cong. (2021).

———. *Developing Core Capabilities for Deep Space Exploration: An Update on NASA's SLS, Orion, and Exploration Ground Systems; Hearings, Day 1, Before Comm. on Science, Space, and Technology*. 116th Cong. (2019).

———. *Discovery on the Frontiers of Space: Exploring NASA's Science Mission; Hearings, Day 1, Before Comm. on Science, Space, and Technology.* 116th Cong. (2019).

——— *Event Horizon Telescope: The Black Hole Seen Round the World; Hearings, Day 1, Before Comm. on Science, Space, and Technology.* 116th Cong. (2019).

———. *The Fiscal Year 2021 Air Force and Space Force Readiness Posture: Hearings, Day 2, Before Comm. on Armed Services.* 116th Cong. (2020).

———. *Keeping Our Sights on Mars: A Review of NASA's Deep Space Exploration Programs and Lunar Proposal; Hearings, Day 1, Before Comm. on Science, Space, and Technology.* 116th Cong. (2019).

———. *The Legacy of Apollo: Hearings, Day 1, Before Comm. on Science, Space, and Technology.* 116th Cong. (2019).

———. *Military Space Operations, Policy, and Programs: Hearings, Day 1, Before Comm. on Armed Services,* 116th Cong. (2019).

———. *Moon Landings to Mars Exploration: The Role of Small Business Innovation in America's Space Program; Hearings, Day 1, Before Comm. on Small Business and Entrepreneurship.* 116th Cong. (2019).

———. *Near-Peer Advancements in Space and Nuclear Weapons: Hearings, Day 1, Before Comm. on Armed Services.* 117th Cong. (2021).

———. *The Proposal to Establish a United States Space Force: Hearings, Day 1, Before Comm. on Armed Services.* 116th Cong. (2019).

———. *Regulating Space: Innovation, Liberty, and International Obligations; Hearings, Day 1, Before Comm. on Science, Space, and Technology.* 115th Cong. (2017).

———. *Reopening the American Frontier: Exploring How the Outer Space Treaty Will Impact American Commerce and Settlement in Space; Hearings, Day 1, Before Comm. on Commerce, Science, and Transportation.* 115th Cong. (2017).

———. *A Review of NASA's Plans for the International Space Station and Future Activities in Low Earth Orbit: Hearings, Day 1, Before Comm. on Science, Space, and Technology.* 116th Cong. (2019).

———. *A Review of the President's Fiscal Year 2022 Budget Proposal for NASA: Hearings, Day 1, Before the Comm. on Science, Space, and Technology.* 117th Cong. (2021).

———. *Space Situational Awareness: Key Issues in an Evolving Landscape; Hearings, Day 2, Before Comm. on Science, Space, and Technology.* 116th Cong. (2020).

———. *Surveying the Space Weather Landscape: Hearings, Day 2, Before Comm. on Science, Space, and Technology*. 115th Cong. (2018).

US Senate. "Frequently Asked Questions About Committees." Accessed November 16, 2022. https://www.senate.gov/committees/committees_faq .htm.

———. "Rule XXVI (d). Committee Procedure." *Standing Rules of the Senate*, revised January 24, 2013. Document 113-18.

Valentine, David. "Exit Strategy: Profit, Cosmology, and the Future of Humans in Space." *Anthropological Quarterly* 84, no. 4 (2012): 1045–1067.

Valentine, David, Valerie Olson, and Debbora Battaglia. "Extreme: Limits and Horizons in the Once and Future Cosmos." *Anthropological Quarterly* 85, no. 4 (2012): 1007–1026.

Vertesi, Janet. *Seeing Like a Rover: Images and Interaction on the Mars Exploration Rover Mission*. Chicago: University of Chicago Press, 2015.

Villafuerte, Paola. "Decolonizing Science: What Is 'Parachute Science?'" Institute for the Future of Education, July 20, 2020. https://observatory.tec.mx/edu -news/what-is-colonial-science.

Vincent, James. "'Goodnight Earth. Goodnight Humanity': China's Jade Rabbit Rover Tweets Its Own Death." *Independent*, February 3, 2014. https://www .independent.co.uk/tech/goodnight-earth-goodnight-humanity-china-s -jade-rabbit-rover-tweets-its-own-death-9103864.html.

Vinsel, Lee. "Prophecy and Politics, or What Are the Uses of the 'Fourth Industrial Revolution?'" Medium, March 25, 2019. https://medium.com /whats-at-stake-in-a-fourth-industrial-revolution/prophecy-and-politics-or -what-are-the-uses-of-the-fourth-industrial-revolution-30710f349ee9.

Vinsel, Lee, and Andy Russell. *The Innovation Delusion: How Our Obsession with the New Has Disrupted the Work That Matters Most*. New York: Currency, 2020.

Virgil. *Aeneid*. Translated by H. R. Fairclough. In Loeb Classical Library, vols. 63 & 64. Cambridge: Harvard University Press, 1916.

Virgin Galactic News. "Virgin Galactic Successfully Completes First Fully Crewed Spaceflight." July 11, 2021. https://investors.virgingalactic.com /news/news-details/2021/Virgin-Galactic-Successfully-Completes-First -Fully-Crewed-Spaceflight/default.aspx.

Virgin Hyperloop. "Smithsonian FUTURES x Virgin Hyperloop." YouTube, December 14, 2021. https://www.youtube.com/watch?v=cm2SuheeGXs (video no longer available).

Voosen, Paul. "Trump White House Quietly Cancels NASA Research Verifying Greenhouse Gas Cuts." *Science*, May 9, 2018. https://www.science.org /content/article/trump-white-house-quietly-cancels-nasa-research -verifying-greenhouse-gas-cuts.

Wall, Mike. "Asteroid-Mining Project Aims for Deep-Space Colonies." Space. com, January 22, 2013. https://www.space.com/19368-asteroid-mining -deep-space-industries.html.

Wells-Jenson, Sheri. "The Case for Disabled Astronauts." *Scientific American*, May 30, 2018. https://blogs.scientificamerican.com/observations/the-case -for-disabled-astronauts/.

White House. "The Biden-Harris Administration Immediate Priorities." Accessed November 2, 2022. https://www.whitehouse.gov/priorities/.

Willow, Anna J., and Kelly A. Yotebieng, eds. *Anthropology and Activism: New Contexts, New Conversations.* New York: Routledge, 2020.

Wilson, Elizabeth. *Affect and Artificial Intelligence.* Seattle: University of Washington Press, 2010.

Wolfe, Patrick. "Settler Colonialism and the Elimination of the Native." *Journal of Genocide Research* 8, no. 4 (2006): 387–409.

Wynne, Brian. "Misunderstood Misunderstanding: Social Identities and Public Uptake of Science." *Public Understanding of Science* 1, no. 3 (1992): 281–304.

Yeager, Sean. "A Disordered Review of Chanda Prescod-Weinstein, *The Disordered Cosmos.*" *Postmodern Culture* 31, no. 3 (2021).

Young, Jane. "Pity the Indians of Outer Space: Native American Views on the Space Program." *Western Folklore* 46, no. 4 (October 1987): 273.

Zevallos, Zuleyka. "Rethinking the Narrative of Mars Colonisation." *Other Sociologist*, March 26, 2015. https://othersociologist.com/2015/03/26 /rethinking-the-narrative-of-mars-colonisation/.

Zuiderent-Jerak, Teun. "Blurring the Center: On the Politics of Ethnography," *Scandinavian Journal of Information Systems* 14, no. 2 (2002): article 9.

———. "Embodied Interventions—Interventions on Bodies: Experiments in Practices of Science and Technology Studies and Hemophilia Care." *Science, Technology & Human Values* 35, no. 5 (2010): 677–710.

# INDEX